DIGITAL SYSTEMS DESIGN
AND PROTOTYPING
USING FIELD PROGRAMMABLE LOGIC

DIGITAL SYSTEMS DESIGN
AND PROTOTYPING
USING FIELD PROGRAMMABLE LOGIC

by

Zoran Salcic
The University of Auckland

and

Asim Smailagic
Carnegie Mellon University

KLUWER ACADEMIC PUBLISHERS
Boston / Dordrecht / London

Distributors for North America:
Kluwer Academic Publishers
101 Philip Drive
Assinippi Park
Norwell, Massachusetts 02061 USA

Distributors for all other countries:
Kluwer Academic Publishers Group
Distribution Centre
Post Office Box 322
3300 AH Dordrecht, THE NETHERLANDS

Library of Congress Cataloging-in-Publication Data

A C.I.P. Catalogue record for this book is available
from the Library of Congress.

Printed on acid-free paper.

Printed in the United States of America

Table Of Contents

x

LIST OF FIGURES

LIST OF TABLES

LIST OF EXAMPLES

To Dushka and Brana,
and
Srdjan, Goran, Vedrana, and Gorana.

Foreword

Design is the major differentiator between products and processes in the marketplace. Design is the creative process that determines the function, form, capacity, and ultimately the utility of a product or process. Decisions during the design phase impact the ease with which the product can be manufactured or a process or building can be operated. By the time that ten percent of the total effort is expended on a new product, over 80% of the decisions have been made. A poor design can not be iterated into a superior product no matter how much down stream effort is expended. While manufacturing technologies have become more efficient there have not been complementary gains in design to the point that design has become one of the three most critical technologies for the U.S. semiconductor industry.

In addition, there is rapid change in the technology of commodity components such as microprocessors and memory chips. Microprocessors are improving performance at 60% per year compounded while memory chip capacity quadruples every three years. These rapid changes in base technology lead to rapid product obsolescence. Major electronics-based companies derive the majority of their revenue from products that are less than three years old. In addition, global competition can require almost continuous introduction of new products. Consumer electronic products often have a design and manufacturing cycle of less than six months. Projecting this trend implies that in the future fully customized products will be fabricated on demand.

Rapid design, prototyping, and manufacturing of complex digital products is fundamental to success in the marketplace. Design is an iterative process and rapid prototyping allows experimentation and evaluation in a timely manner. This is a ground-breaking book that bridges the gap between digital design theory and practice. It provides a unifying terminology for describing FPLD technology. In addition to introducing the technology it also describes the design methodology and tools required to harness this technology. It introduces two hardware description languages (e.g., AHDL and VHDL). Design is best learned by practice and the book supports this notion with abundant case studies.

But the book has value not only for students but also the practicing engineer. An important solution to incorporating advanced technology into products in a timely manner is the use of mass-produced, general-purpose components such as microprocessors and memories "glued" together by Field-Programmable Logic Devices (FPLD). In addition to providing the interfaces between mass production chips, FPLDs can provide special-purpose input/output functionality for product customization. FPLDs are a key enabling technology for customized products to keep pace with the rapid rate of change of component technology. The book illustrates how FPLDs can be used to "glue together" microprocessors, memories, and control chips to rapidly realize a novel electronic system.

In summary, in the future products and processes will be even more individualized with end users playing a key role in the specification of the product. The design process will draw upon expertise and multiple disciplines to generate an effective solution. In the creation of new products and processes, industry must deliver "customized designs with the economy and high quality previously achieved only in mass production." FPLDs is one of those technologies that will make this revolution possible and this book shows how to harness the potential of FPLDs in practical designs.

Daniel P. Siewiorek

PREFACE

This book focuses on digital systems design and Field- Programmable Logic Devices (FPLDs), combining them into an entity useful for designers in the areas of digital systems and rapid system prototyping. It is also useful for the growing community of engineers and researchers dealing with the exciting field of FPLDs, reconfigurable, and programmable logic. Our goal is to bring these areas to the students studying digital system design, computer design, and related topics, as to show how very complex circuits can be implemented at the desk. Hardware and software designers are getting closer every day by the emerging technologies of in-circuit reconfigurable and in-system programmable logic of very high complexity.

Field-programmable logic has been available for a number of years. The role of FPLDs has evolved from simply implementing the system "glue-logic" to the ability to implement very complex system functions, such as microprocessors and microcomputers. The speed with which these devices can be programmed makes them ideal for prototyping and education. Low production cost makes them competitive for small to medium volume productions. These devices make possible new sophisticated applications, and bring-up new hardware/software trade-offs and diminish the traditional hardware/software demarcation line. Advanced design tools are being developed for automatic compilation of complex designs and routings to custom circuits.

To our knowledge, this book makes a pioneering effort to present rapid prototyping and generation of computer systems using FPLDs. Rapid prototyping systems composed of programmable components show great potential for full implementation of microelectronics designs. Prototyping systems based on field- programmable devices present many technical challenges affecting system utilization and performance.

The book contains eight chapters. Chapter 1 represents an introduction into the Field-Programmable Logic. Main types of FPLDs are introduced, including programming technologies, logic cell architectures, and routing architectures used to interconnect logic cells. Architectural features are discussed to allow the reader to compare different devices appearing on the market, sometimes using confusing terminology and hiding the real nature of the devices. Also, the main characteristics of the design process using FPLDs are discussed and the differences to the design for custom integrated circuits underlined. The necessity to introduce and use new advanced tools when designing complex digital systems is also emphasized.

Chapter 2 describes the field-programmable devices of the two major manufacturers in the market, Altera and Xilinx. It does not mean that devices from other manufacturers are inferior to presented ones. The purpose of this book is not to compare different devices, but to emphasize the most important features found in the majority of FPLDs, and their use in complex digital system prototyping and design. Altera and Xilinx invented some of the concepts found in major types of field-programmable logic and also produce devices which employ all major programming technologies. Complex Programmable Logic Devices (CPLDs) and Field-Programmable Gate Arrays (FPGAs) are presented in Chapter 2, along with their main architectural and application-oriented features. Although sometimes we use different names to distinguish CPLDs and FPGAs, usually with the term FPLD we will refer to both types of devices.

Chapter 3 covers aspects of the design methodology and design tools used to design with FPLDs. The need for tightly coupled design frameworks, or environments, is discussed and the hierarchical nature of digital systems design. All major design description (entry) tools are introduced including schematic entry tools and hardware description languages. The complete design procedure, which includes design entry, processing, and verification, is shown in an example of a simple digital system. An integrated design environment for FPLD-based designs, the Altera's Max+PLUS II environment, is introduced. It includes various design entry, processing, and verification tools.

Chapter 4 is devoted to the design using Altera's Hardware Description Language (AHDL). First, the basic features of AHDL are introduced without a formal presentation of the language. Small examples are used to illustrate its features and how they are used. The design of combinatorial logic in AHDL including the implementation of bidirectional pins, standard sequential circuits such as registers and counters, and state machines is presented. Vendor supplied and user defined macrofunctions appear as a library entities. The implementation of user designs as hierarchical projects consisting of a number of subdesigns is also shown. AHDL, as a lower level hardware description language, allows user control of resource assignments and very effective control of the design fit to target either speed or size optimization. Still, the designs specified in AHDL can be of behavioral or structural type and easily retargeted, without change, to another device without the need for the change of the design specification.

Chapter 5 shows how designs can be handled using primarily AHDL, but also in the combination with the more convenient schematic entry tools. Two design case studies, which include a number of combinational and sequential circuit designs are shown in this chapter. The first example is an electronic lock which consists of a hexadecimal keypad as the basic input device and a number of LEDs as the output indicators of different states. The lock activates an unlock signal after recognizing the input of a sequence of five digits acting as a kind of password. The second example is a temperature control circuitry which enables temperature control in a small chamber (incubator). The temperature controller continuously scans the current temperature and activates one of two actuators, a lamp for heating or a fan for cooling. The controller allows set up of a low and high temperature limits range where the current temperature should be maintained. It also provides the basic interface with the operator in the form of hexadecimal keypad as input and 7-segment display and couple of LEDs as output. Both designs fit into the standard Altera's devices.

Chapter 6 includes a more complex example of a simple custom configurable microprocessor called SimP. The microprocessor contains a fixed core that implements a set of instructions and addressing modes which serve as the base for more complex microprocessors with additional instructions and processing capabilities as needed by a user and/or application. It provides the mechanisms to be extended by the designers in various directions and with some further modifications it can be converted to become a sort of dynamically reconfigurable processor. Most of the design is specified in AHDL to demonstrate the power of the language. A performance analysis of the design has also been shown.

Chapter 7 provides an introduction to VHDL as a more abstract and powerful hardware description language, which is also accepted as an IEEE standard. The goal of this chapter is to demonstrate how VHDL can be used in digital system design. A subset of the language features is used to provide designs that can almost always be synthesized. The features of sequential and concurrent statements, objects, entities, architectures, and configurations, allow very abstract approaches to system design, at the same time controlling design in terms of versions, reusability, or exchangeability of the portions of design. Combined with the flexibility and potential reconfigurability of FPLDs, VHDL represents a tool which will be more and more in use in digital system prototyping and design. The example of an input sequence classifier and recognizer is used to demonstrate the use of VHDL in digital systems design that are easily implemented in FPLDs.

Finally Chapter 8 is used to present a case study of a digital system based on the combination of a standard microprocessor and FPLD implemented logic. The VuMan wearable computer, developed at Carnegie Mellon University (CMU), is presented in this chapter. Examples of the VuMan include the design of interfacing logic and various peripheral controllers, for which FPLDs are used as the most appropriate prototyping and implementation technology.

The book is based on lectures we have given in different courses and advanced research projects at Auckland University and CMU, various projects carried out in the course of different degrees, and the courses for professional engineers who are entering the field of FPLDs and CAD tools for complex digital systems design. As with any book, it is still open and can be improved and enriched with new materials, especially due to the fact that the subject area is rapidly changing. The complete Chapter 8 represents a portion of the VuMan project carried out at Carnegie Mellon University. Some of the original VuMan designs are modified for the purpose of this book at Auckland University.

A special gratitude is directed to the Altera Corporation for enabling us to try many of the concepts using their tools and devices in the course of its University Program Grant. Also Altera made possible the opportunity for numerous students at Auckland University to take part in various courses designing digital systems using these new technologies. A special thanks to Altera for providing design software on CD ROM included with this book. This book would not be possible without the supportive environment at Auckland University, New Zealand, and Carnegie Mellon University, USA, as well as early support from Cambridge University, Czech Technical University, University of Edinburgh, and Sarajevo University where we spent memorable years teaching and conducting research.

Z. A. Salcic   A. Smailagic
Auckland, New Zealand   Pittsburgh, USA

About the Accompanying CD-ROM

Digital Systems Design and Prototyping Using Field Programmable Logic, First Edition includes a CD-ROM that contains Altera's MAX+PLUS II 7.21 Student Edition programmable logic development software. MAX+PLUS II is a fully integrated design environment that offers unmatched flexibility and performance. The intuitive graphical interface is complemented by complete and instantly accessible on-line documentation, which makes learning and using MAX+PLUS II quick and easy. MAX+PLUS II version 7.21 Student Edition offers the following features:

✓ Operates on PCs running Windows 3.1, Windows 95, and Windows NT 3.51 and 4.0.
✓ Graphical and text-based design entry, including Altera Hardware Description Language (AHDL) and VHDL.
✓ Design compilation for product-term (MAX 7000S) and look-up table (FLEX 10K) device architectures.
✓ Design verification with full timing simulation.

Installing with Windows 3.1 and Windows NT 3.51

Insert the MAX+PLUS II CD-ROM in your CD-ROM drive. In the Windows Program Manager, choose **Run** and type: *<CD-ROM drive>*: \pc\maxplus2\install in the *Command Line* box. You are guided through the installation procedure.

Installing with Windows 95 and Windows NT 4.0

Insert the MAX+PLUS II CD-ROM in your CD-ROM drive. In the Start menu, choose **Run** and type: *<CD-ROM drive>*: \pc\maxplus2\install in the *Open* box. You are guided through the installation procedure.

Registration & Additional Information

To register and obtain an authorization code to use the MAX+PLUS II software, to **http://www.altera.com/maxplus2-student**. For complete installation instructions, refer to the **read.me** file on the CD-ROM or to the *MAX+PLUS II Getting Started Manual*, available on the Altera world-wide web site (**http://www.altera.com**).

This CD-ROM is distributed by Kluwer Academic Publishers with *ABSOLUTELY NO SUPPORT* and *NO WARRANTY* from Kluwer Academic Publishers.

Kluwer Academic Publishers shall not be liable for damages in connection with, or arising out of, the furnishing, performance or use of this CD-ROM.

1 INTRODUCTION TO FIELD PROGRAMMABLE LOGIC DEVICES

Programmable logic design is beginning the same paradigm shift that drove the success of logic synthesis within ASIC design, namely the move from schematics to HDL based design tools and methodologies. Technology advancements, such as 0.5 micron triple level metal processing and architectural innovations such as large amount of on-chip memory, have significantly broadened the applications for Field-Programmable Logic Devices (FPLDs).

This chapter represents an introduction to the Field-Programmable Logic. The main types of FPLDs are introduced, including programming technologies, logic cell architectures, and routing architectures used to interconnect logic cells. Architectural features are discussed to allow the reader to compare different devices appearing on the market. The main characteristics of the design process using FPLDs are also discussed and the differences to the design for custom integrated circuits underlined. In addition, the necessity to introduce and use new advanced tools when designing complex digital systems is emphasized.

1.1. Introduction

FPLDs represent a relatively new development in the field of VLSI circuits. They implement thousands of logic gates in multilevel structures. The architecture of an FPLD, similar to that of a Mask-Programmable Logic Device (MPLD), consists of an array of logic cells that can be interconnected by programming to implement different designs. The major difference between an FPLD and an MPLD is that an MPLD is programmed using integrated circuit fabrication to form metal interconnections while an FPLD is programmed using electrically programmable switches similar to ones in traditional Programmable Logic Devices (PLDs). FPLDs can achieve much higher levels of integration than traditional PLDs due to their more complex routing architectures and logic implementation. PLD routing architectures are very simple with inefficient crossbar like structures in which every output is connectable to every input through one switch. PLD logic is implemented using AND-OR logic with wide input AND gates. FPLD routing architectures provide a more efficient MPLD-like routing where each connection typically passes through several switches. FPLD logic is implemented using multiple levels of lower fan-in gates which is often more compact than two-level implementations.

An FPLD manufacturer makes a single, standard device that users program to carry out desired functions. Field programmability comes at a cost in logic density and performance. FPLD capacity trails MPLD capacity by about a factor of 10 and FPLD performance trails MPLD performance by about a factor of three. Why then FPLDs? FPLDs can be programmed in seconds rather than weeks, minutes rather than the months required for production of mask-programmed parts. Programming is done by end users at their site with no IC masking steps. FPLDs are currently available in densities up to 100,000 gates in a single device. This size is large enough to implement many digital systems on a single chip and larger systems can be implemented on multiple FPLDs on the standard PCB or in the form of Multi-Chip Modules (MCM). Although the unit costs of an FPLD is higher than an MPLD of the same density, there is no up-front engineering charges to use an FPLD, so they are more cost-effective for many applications. The result is a low-risk design style, where the price of logic error is small, both in money and project delay.

FPLDs are useful for rapid product development and prototyping. They provide very fast design cycles, and, in the case that the major value of the product is in algorithms or fast time-to-market they prove to be even cost-effective as the final deliverable product. Since FPLDs are fully tested after manufacture, user designs do not require test program generation, automatic test pattern generation, and design for testability. Some FPLDs have found a suitable place in designs that require reconfiguration of the hardware structure during system operation, functionality can change "on the fly."

An illustration of device options ratings, that include standard discrete logic, FPLDs, and custom logic is given in Figure 1.1. Although not quantitative, the figure demonstrates many advantages of FPLDs over other types of available logic.

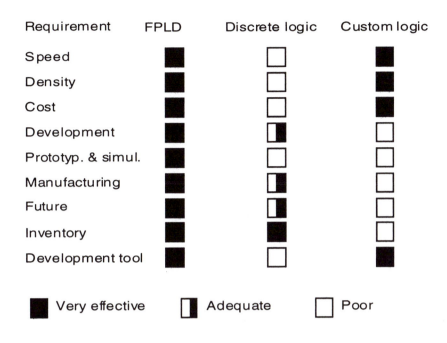

Requirement	FPLD	Discrete logic	Custom logic
Speed	■	□	■
Density	■	□	■
Cost	■	□	■
Development	■	▊	□
Prototyp. & simul.	■	□	□
Manufacturing	■	▊	□
Future	■	▊	□
Inventory	■	■	□
Development tool	■	□	■

■ Very effective	▊ Adequate	□ Poor

Figure 1.1 Device options ratings for different device technologies

The purpose of Figure 1.1 and this discussion is to point out some of the major features of currently used options for digital system design, and show why we consider FPLDs as the most promising technology for implementation of a very large number of digital systems.

Until recently only two major options were available to digital system designers.

- First, they could use Small-Scale Integrated (SSI) and Medium-Scale Integrated (MSI) circuits to implement a relatively small amount of logic with a large number of devices.

- Second, they could use a Masked-Programmed Gate Array (MPLD) or simply gate array to implement tens or hundreds of thousands of logic gates on a single integrated circuit in multi-level logic with wiring between logic levels. The wiring of logic is built during the manufacturing process requiring a custom mask for the wiring. The low volume MPLDs have been expensive due to high mask-making charges.

As intermediate solutions for the period during the 1980s and early 1990s various kinds of simple Programmable Logic Devices (PLDs) were available. A simple PLD is a general purpose logic device capable implementing the logic of tens or hundreds of SSI circuits and customize logic functions in the field using inexpensive programming hardware. Large designs require a multi-level logic implementation introducing high power consumption and large delays.

Field-Programmable Logic Devices offer the benefits of both PLDs and MPLDs. They allow the implementation of thousands of logic gates in a single circuit and can be programmed by designers on the site not requiring expensive manufacturing processes. The discussion below is largely targeted to a comparison of FPLDs and MPLDs as the technologies suitable for complex digital system design and implementation.

1.1.1 Speed

FPLDs offer devices which operate at speeds approaching 200 MHz in many applications. Obviously, speeds are higher than in systems implemented by SSI circuits, but lower than the speeds of MPLDs. The main reason for this comes from the FPLD programmability. Programmable interconnect points add resistance to the internal path, while programming points in the interconnect mechanism add capacitance to the internal path. Despite these disadvantages when compared to MPLDs, FPLD speed is adequate for most applications. Also, some dedicated architectural features of FPLDs can eliminate unneeded programmability in speed critical paths.

By moving FPLDs to faster processes, application speed can be increased by simply buying and using a faster device without design modification. The situation with MPLDs is quite different; new processes require new mask-making and increase the overall product cost.

1.1.2 Density

FPLD programmability introduces on-chip programming overhead circuitry requiring area that cannot be used by designers. As a result, the same amount of logic for FPLDs will always be larger and more expensive than MPLDs. However, a large area of the die cannot be used for core functions in MPLDs due to the I/O pad limitations. The use of this wasted area for field programmability does not result in an increase of area for the resulting FPLD. Thus, for a given number of gates, the size of an MPLD and FPLD is dictated by the I/O count so the FPLD and MPLD capacity will be the same. This is especially true with the migration of FPLDs to submicron processes. MPLD manufacturers have already shifted to high-density products leaving designs with less than 20,000 gates to FPLDs.

1.1.3 Development Time

FPLD development is followed by the development of tools for system designs. All those tools belong to high-level tools affordable even to very small design houses. The development time primarily includes prototyping and simulation while the other phases, including time-consuming test pattern generation, mask-making, wafer fabrication, packaging, and testing are completely avoided. This leads to the typical development times for FPLD designs measured in days or weeks, in contrast to MPLD development times in several weeks or months.

1.1.4 Prototyping and Simulation Time

While the MPLD manufacturing process takes weeks or months from design completion to the delivery of finished parts, FPLDs require only design completion. Modifications to correct a design flaw are quickly and easily done providing a short turn around time that leads to faster product development and shorter time-to-market for new FPLD-based products.

Proper verification requires MPLD users to verify their designs by extensive simulation before manufacture introducing all of the drawbacks of the speed/accuracy trade-off connected with any simulation. In contrast, FPLDs simulations are much simpler due to the fact that timing characteristics and models are known in advance. Also, many designers avoid simulation completely and choose in-circuit verification. They implement the design and use a functioning part as a prototype which operates at full speed and absolute time accuracy. A prototype can be easily changed and reinserted into the system within minutes or hours.

FPLDs provide low-cost prototyping, while MPLDs provide low-cost volume production. This leads to prototyping on the FPLD and then switching to an MPLD for volume production. Usually there is no need for design modification when retargeting to an MPLD, except sometimes when timing path verification fails. Some FPLD vendors offer mask-programmed versions of their FPLDs giving users flexibility and advantages of both implementation methods.

1.1.5 Manufacturing time

All integrated circuits must be tested to verify manufacturing and packaging. The test is different for each design. MPLDs typically incur three types of costs associated with testing.

- on-chip logic to enable easier testing
- generation of test programs for each design
- testing the parts when manufacturing is complete

Because they have a simple and repeatable structure, the test program for one FPLD device is same for all designs and all users of that part. It further justifies all reasonable efforts and investments to produce extensive and high quality test programs that will be used during the lifetime of the FPLD. Users are not required to write design specific tests because manufacturer testing verifies that every FPLD will function for all possible designs implemented. The consequences of manufacturing chips from both categories are obvious. Once verified, FPLDs can be manufactured in any quantity and delivered as fully tested parts ready for design implementation while MPLDs require separate production preparation for each new design.

1.1.6 Future modifications

Instead of customizing the part in the manufacturing process as for MPLDs, FPLDs are customized by electrical modifications. The electrical customization takes milliseconds or minutes and can even be performed without special devices, or with low cost programming devices. Even more, it can usually be performed in-system, meaning that the part can already be on the printed circuit board reducing the danger of the damage due to uncareful handling. On the other hand, every modified design to be implemented in an MPLD requires a custom mask that costs several thousands dollars that can only be amortized over the total number of units manufactured.

1.1.7 Inventory risk

An important feature of FPLDs is low inventory risk, similar to SSI and MSI parts. Since actual manufacturing is done at the time of programming a device, the same part can be used for different functionality and different designs. This is not found in an MPLD since the functionality and application is fixed forever once it is produced. Also, the decision on the volume of MPLDs must be made well in advance of the delivery date, requiring concern with the probability that too many or not enough parts are ordered to manufacture. Generally, FPLDs are connected with very low risk design in terms of both money and delays. Rapid and easy prototyping enables all errors to be corrected with short delays, but also gives designers the chance to try more risky logic designs in the early stages of product development. Development tools used for FPLD designs usually integrate the whole range of design entry, processing, and simulation tools which enable easy reusability of all parts of a correct design.

FPLD designs can be made with the same design entry tools used in traditional MPLDs and Application Specific Integrated Circuits (ASICs) development. The resulting netlist is further manipulated by FPLD specific fitting, placement, and routing algorithms that are available either from FPLD manufacturers or CAE vendors. However, FPLDs also allow designing on the very low device dependent level providing the best device utilization, if needed.

1.1.8 Cost

Finally, the above introduced options reflect on the costs. The major benefit of an MPLD-based design is low cost in large quantities. The actual volume of the products determines which technology is more appropriate to be used. FPLDs have much lower costs of design development and modification, including initial Non-Recurring Engineering (NRE) charges, tooling, and testing costs. However, larger die area and lower circuit density result in higher manufacturing costs per unit. The break-even point depends on the application and volume, and is usually at between ten and twenty thousand units for large capacity FPLDs. This limit is even higher when an integrated volume production approach is applied, using a combination of FPLDs and their corresponding masked-programmed counterparts. Integrated volume production also introduces further flexibility, satisfying short term needs with FPLDs and long term needs at the volume level with masked-programmed devices.

1.2 Types of FPLDs

The general architecture of an FPLD is shown in Figure 1.2. A typical FPLD consists of a number of logic cells that are used for logic functions. Logic cells are arranged in a form of a matrix. Interconnection resources connect logic cell outputs and inputs, as well as input/output blocks used to connect FPLD with the outer world.

Despite the same general structure, concrete implementations of FPLDs differ among the major components. There is a difference in approach to circuit programmability, internal logic cell structure, and routing mechanisms.

An FPLD logic cell can be a simple transistor or a complex microprocessor. Typically, it is capable of implementing combinational and sequential logic functions of different complexities.

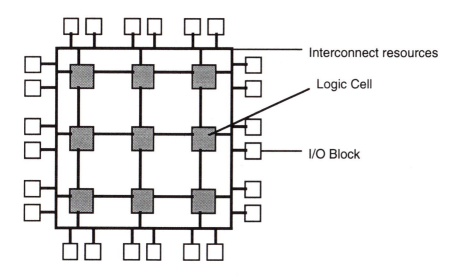

Figure 1.2 FPLD architecture

Current commercial FPLDs employ logic cells that are based on one or more of the following:

- Transistor pairs

- Basic small gates, such as two-input NANDs or XORs

- Multiplexers

- Look-up tables (LUTs)

- Wide-fan-in AND-OR structures

Three major programming technologies, each associated with area and performance costs, are commonly used to implement the programmable switch for FPLDs. These are:

- Static Random Access Memory (SRAM), where the switch is a pass transistor controlled by the state of a SRAM bit

- EPROM, where the switch is a floating-gate transistor that can be turned off by injecting charge onto its floating gate, and

- Antifuse, which, when electrically programmed, forms a low resistance path.

In all cases, a programmable switch occupies a larger area and exhibits much higher parasitic resistance and capacitance than a typical contact used in a custom MPLDs. Additional area is also required for programming circuitry, resulting in higher density and lower speed of FPLDs compared to MPLDs.

An FPLD routing architecture incorporates wire segments of varying lengths which can be interconnected with electrically programmable switches. The density achieved by an FPLD depends on the number of wires incorporated. If the number of wire segments is insufficient, only a small fraction of the logic cells can be utilized. An excessive number of wire segments wastes area. The distribution of wire segments greatly affects both density and performance of an FPLD. For example, if all segments stretch over the entire length of the device (so called long segments), implementing local interconnections costs area and time. On the other hand, employment of only short segments requires long interconnections to be implemented using many switches in series, resulting in unacceptably large delays.

Both density and performance can be optimized by choosing the appropriate granularity and functionality of logic cell, as well as designing the routing architecture to achieve a high degree of routability while minimizing the number of switches. Various combinations of programming technology, logic cell architecture, and routing mechanisms lead to various designs suitable for specific applications. A more detailed presentation of all major components of FPLD architectures is given in the chapters that follow.

If programming technology and device architecture are combined, three major categories of FPLDs are distinguished:

- Static RAM Field Programmable Logic Arrays, or SRAM FPGAs,

- Complex Programmable Logic Device CPLDs,

- Antifuse FPGAs

Here the major features of these three categories of FPLDs will be presented:

1.2.1 CPLDs

A typical CPLD architecture is shown in Figure 1.3. The user creates logic interconnections by programming EPROM or EEPROM transistors to form wide fan-in gates.

Function Blocks (FBs) are similar to a simple two-level PLD. Each FB contains a PLD AND-array that feeds its macrocells. The AND-array consists of a number of product terms. The user programs the AND-array by turning on EPROM transistors that allow selected inputs to be included in a product term.

A macrocell includes an OR gate to complete AND-OR logic and may also include registers and an I/O pad. It can also contain additional EPROM cells to control multiplexers that select a registered or non-registered output and decide whether or not the macrocell result is output on the I/O pad at that location. Macrocell outputs are connected as additional FB inputs or as the inputs to a global universal interconnect mechanism (UIM) that reaches all FBs on the chip.

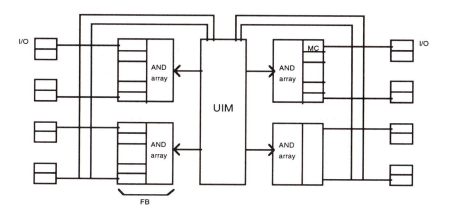

Figure 1.3 Typical CPLD architecture

FBs, macrocells, and interconnect mechanisms vary from one product to another, giving a range of device capacities and speeds.

1.2.2 Static RAM FPGAs

In SRAM FPGAs, static memory cells hold the program. SRAM FPGAs implement logic as lookup tables (LUTs) made from memory cells with function inputs controlling the address lines. Each LUT of 2^n memory cells implements any function of n inputs. One or more LUTs, combined with flip-flops, form a configurable logic block (CLB). CLBs are arranged in a two-dimensional array with interconnect segments in channels as shown in Figure 1.4.

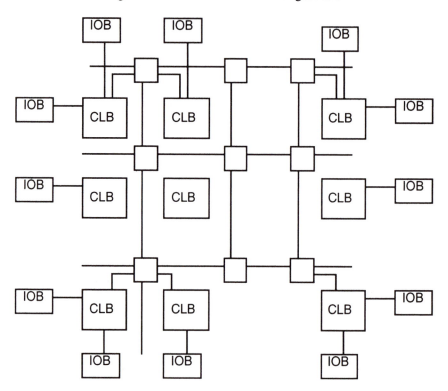

Figure 1.4 Typical SRAM FPGA architecture

Interconnect segments connect to CLB pins in the channels and to the other segments in the switch boxes through pass transistors controlled by configuration memory cells. The switch boxes, because of their high complexity, are not full crossbar switches.

An SRAM FPGA program consists of a single long program word. On-chip circuitry loads this word, reading it serially out of an external memory every time power is applied to the chip. The program bits set the values of all configuration memory cells on the chip, thus setting the lookup table values and selecting which segments connect each to the other. SRAM FPGAs are inherently reprogrammable. They can be easily updated providing designers with new capabilities such as reconfigurability.

1.2.3 Antifuse FPGAs

An antifuse is a two-terminal device that, when exposed to a very high voltage, forms a permanent short circuit (opposite to a fuse) between the nodes on either side. Individual antifuses are small, enabling an antifuse-based architecture to have thousands or millions of antifuses. Antifuse FPGA, as illustrated in Figure 1.5, usually consists of rows of configurable logic elements with interconnect channels between them, much like traditional gate arrays.

Logic Blocks

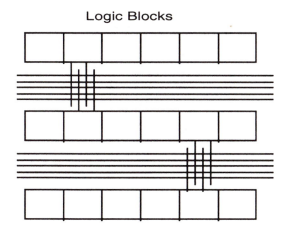

Figure 1.5 Antifuse FPGA architecture

The pins on logic blocks (LBs) extend into the channel. An LB is usually a simple gate-level network, which the user programs by connecting its input pins to fixed values or to interconnect nets. There are antifuses at every wire-to-pin intersection point in the channel and at all wire-to-wire intersection points where channels intersect.

Commercial FPLDs use different programming technologies, different logic cell architectures, and different structures of their routing architectures. A survey of major commercial architectures is given in the rest of this part, and a more detailed presentation of two of FPLD families, Xilinx 3000 and 4000, and Altera 7000 and 8000, is given in Part 2. The majority of design examples are illustrated using Altera's FPLDs.

1.3 Programming Technologies

An FPLD is programmed using electrically programmable switches. The properties of these programmable switches, such as size, volatility, process technology, on-resistance, and capacitance determine the major features of an FPLD architecture. In this section we introduce the most commonly used programmable switch technologies.

1.3.1 SRAM Programming Technology

SRAM programming technology uses Static RAM cells to configure logic and control intersections and paths for signal routing. The configuration is done by controlling pass gates or multiplexers as it is illustrated in Figure 1.6. When a "1" is stored in the SRAM cell in Figure 1.6(a), the pass gate acts as a closed switch and can be used to make a connection between two wire segments. For the multiplexer, the state of the SRAM cells connected to the select lines controls which one of the multiplexers inputs are connected to the output, as shown in Figure 1.6(b). Reprogrammability allows the circuit manufacturer to test all paths in the FPGA by reprogramming it on the tester. The users get well tested parts and 100% "programming yield" with no design specific test patterns and no "design for testability." Since on-chip programming is done with memory cells, the programming of the part can be done an unlimited number of times. This allows prototyping to proceed iteratively, re-using the same chip for new design iterations. Reprogrammability has advantages in systems as well. In cases where parts of the logic in a system are not needed simultaneously, they can be implemented in the same reprogrammable FPGA and FPGA logic can be switched between applications.

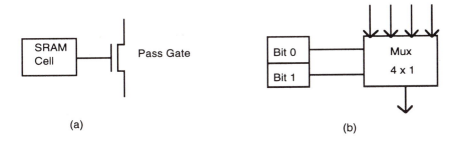

(a)   (b)

Figure 1.6 SRAM Programming Technology

Besides volatility, a major disadvantage of SRAM programming technology is its large area. At least five transistors are needed to implement an SRAM cell, plus at least one transistor to implement a programmable switch. A typical five-transistor memory cell is illustrated in Figure 1.7. There is no separate RAM area on the chip. The memory cells are distributed among the logic elements they control. Since FPGA memories do not change during normal operation, they are built for stability and density rather than speed. However, SRAM programming technology has two further major advantages; fast-reprogrammability and that it requires only standard integrated circuit process technology.

Since SRAM is volatile, the FPGA must be loaded and configured at the time of chip power-up. This requires external permanent memory to provide the programming bitstream such as PROM, EPROM, EEPROM or magnetic disk. This is the reason that SRAM-programmable FPGAs include logic to sense power-on and to initialize themselves automatically, provided the application can wait the tens of milliseconds required to program the device.

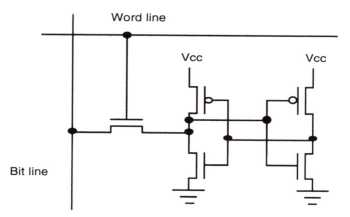

Figure 1.7 Five-transistor Memory Cell

1.3.2 Floating Gate Programming Technology

Floating gate programming technology uses the technology of ultraviolet-erasable EPROM and electrically erasable EEPROM devices. The programmable switch, as shown in Figure 1.8, is a transistor that can be permanently "disabled." To disable the transistor, a charge is injected on the floating polysilicon gate using a high voltage between the control gate and the drain of the transistor. This charge increases the threshold voltage of the transistor so it turns off. The charge is removed by exposing the floating gate to ultraviolet light. This lowers the threshold voltage of the transistor and makes the transistor function normally. Rather than using an EPROM transistor directly as a programmable switch, the unprogrammed transistor is used to pull down a "bit line" when the "word line" is set to high. While this approach can be simply used to provide connection between word and bit lines, it can also be used to implement a wired-AND style of logic, in that way providing both logic and routing.

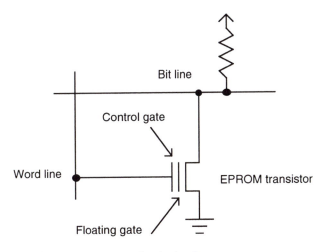

Figure 1.8 Floating gate programming technology

The major advantage of the EPROM programming technology is its reprogrammability. An advantage over SRAM is that no external permanent memory is needed to program a chip on power-on. On the other hand, reconfiguration itself can not be done as fast as in SRAM technology devices. Additional disadvantages are that EPROM technology requires three more processing steps over an ordinary CMOS process, the high on-resistance of an EPROM transistor, and the high static power consumption due to the pull-up resistor used.

EEPROM technology used in some devices is similar to the EPROM approach, except that removal of the gate charge can be done electrically, in-circuit, without ultraviolet light. This gives an advantage of easy reprogrammability, but requires more space due to the fact that EEPROM cell is roughly twice the size of an EPROM cell.

1.3.3 Antifuse Programming Technology

An antifuse is an electrically programmable two-terminal device. It irreversibly changes from high resistance to low resistance when a programming voltage (in excess of normal signal levels) is applied across its terminals. Antifuses offer several unique features for FPGAs , most notably a relatively low on-resistance of 100-600 Ohms and a small size. The layout area of an antifuse cell is generally smaller than the pitch of the metal lines it connects; it is about the same size as a via connecting metal lines in an MPLD. When high voltage (11 to 20 Volts) is applied across its terminals, the antifuse will "blow" and create a low resistance link. This link is permanent. Antifuses are built either using an Oxygen-Nitrogen-Oxygen (ONO) dielectric between an N+ diffusion and polysilicon, or amorphous silicon between metal layers or between polysilicon and the first layer of metal.

Programming an antifuse requires extra circuitry to deliver the high programming voltage and a relatively high current of 5 mA or more. This is done through large transistors to provide addressing to each antifuse.

Antifuses are normally "off" devices. Only a small fraction of the total that need to be turned on must be programmed (about 2% for a typical application). So, other things being equal, programming is faster with antifuses than with "normally on" devices.

Antifuse reliability must be considered for both the unprogrammed and programmed states. Time dependent dielectric breakdown (TDDB) reliability over 40 years is an important consideration. It is equally important that the resistance of a programmed antifuse remains low during the life of the part. Analysis of ONO dielectrics shows that they do not increase the resistance with time. Additionally, the parasitic capacitance of an unprogrammed amorphous antifuse is significantly lower than for other programming technologies.

1.3.4 Summary of Programming Technologies

Major properties of each of above presented programming technologies are shown in Table 1.1. All data assumes a 1.2 μm CMOS process technology. The most recent devices use much higher density devices and many of them are implemented in 0.6 or even 0.5 μm CMOS process technology.

Table 1.1 Comparison of Programming technologies

Technology and Process	Volatile ?	Reprogram mability ?	Area	R (ohm) (on switch)	C (fF) (parasitic)	# Extra fabric. steps
SRAM Mux Pass Trans. 1.2 μm CMOS	Yes	Yes In-circuit	Large	0.5-2K	10-20	0
ONO Antifuse 1.2 μm CMOS	No	No	Fuse small Prog. trans. large	300-600	5	3
Amor-phous Antifuse 1.2 μm CMOS	No	No	Fuse small Prog. trans. large	50-100	1.1-1.3	3
EPROM 1.2 μm CMOS	No	Yes Out of circuit	Small in array	2-4K	10-20	3
EEPROM 1.2 μm CMOS	No	Yes In-circuit	2 x EPROM	2-4K	10-20	>5

1.4. Logic Cell Architecture

In this section we present a survey of commercial FPLD logic cell architectures in use today, including their combinational and sequential portions. FPLD logic cells differ both in size and implementation capability. A two transistor logic cell can only implement a small size inverter, while the look-up table logic cells can implement any logic function of several input variables and is significantly larger. To capture these differences we usually classify logic blocks by their granularity. Since granularity can be defined in various ways (as the number of Boolean functions that the logic block can implement, the number of two-input AND gates, total number of transistors, etc.), we choose to classify commercial blocks into just two categories: fine-grain and coarse-grain.

Fine-grain logic cells resemble MPLD basic cells. The most fine grain logic cell would be identical to a basic cell of an MPLD and would consist of few transistors that can be programmably interconnected.

The FPGA from Crosspoint Solutions uses a single transistor pair in the logic cell. In addition to the transistor pair tiles, as depicted in Figure 1.9, the cross-point FPGA has a second type of logic cell, called a RAM logic tile, that is tuned for the implementation of random access memory, but can also be used to build other logic functions.

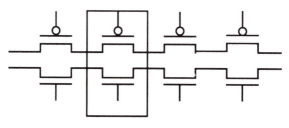

Transistor Pair

Figure 1.9 Transistor pair tiles in cross-point FPGA

A second example of a fine-grain FPGA architecture is the FPGA from Plessey. Here the basic cell is a two-input NAND gate as illustrated in Figure 1.10. Logic is formed in the usual way by connecting the NAND gates to implement the desired function. If the latch is not needed, then the configuration memory is set to make the latch permanently transparent.

Several other commercial FPGAs employ fine-grain logic cells. The main advantage of using fine-grain logic cells is that the usable cells are fully utilized. This is because it is easier to use small logic gates efficiently and the logic synthesis techniques for such cells are very similar to those for conventional MPGAs (Mask- Programmable Gate Arrays) and standard cells.

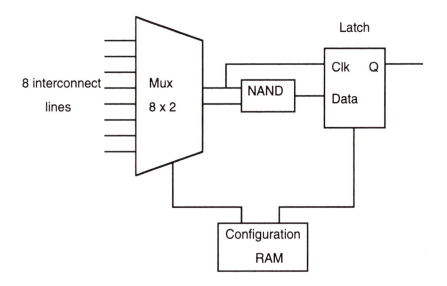

Figure 1.10 The Plessey Logic Cell

The main disadvantage of fine-grain cells is that they require a relatively large number of wire segments and programmable switches. Such routing resources are costly in both delay and area. If a function that could be packed into a few complex cells must instead be distributed among many simple cells, more connections must be made through the programmable routing network. As a result, FPGAs with fine-grain logic cells are in general slower and achieve lower densities than those using coarse-grain logic cells.

As a rule of thumb, an FPGA should be as fine-grained as possible while maintaining good routability and routing delay for the given switch technology. The cell should be chosen to implement a wide variety of functions efficiently, yet have minimum layout area and delay.

Actel's logic cells have been designed on the base of usage analysis of various logic functions in actual gate array applications. The Act-1 family uses one general-purpose logic cell as shown in Figure 1.11. The cell is composed of three 2-to-1 multiplexers, one OR gate, 8 inputs, and one output. Various macrofunctions (AND, NOR, flip-flops, etc.) can be implemented by applying each input signal to the appropriate cell inputs and tying other cell inputs to 0 or 1. The cell can implement all combinational functions of two inputs, all functions of three inputs with at least one positive input, many functions of four inputs, and some ranging up to eight inputs. Any sequential macro can be implemented from one or more cells using appropriate feedback routings.

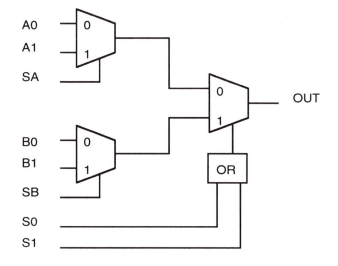

Figure 1.11 Act-1 logic cell

Further analysis of macros indicate that a significant proportion of the nets driving the data input of flip-flop have no other fan-out. This motivated the use of a mixture of two specialized cells for Act-2 and Act-3 families. The "C" cell and its equivalent shown in Figure 1.12 are modified versions of the Act-1 cell re-optimized to better accommodate high fan in combinational macros. It actually represents a 4-to-1 multiplexer and two gates, implementing a total of 766 distinct combinational functions.

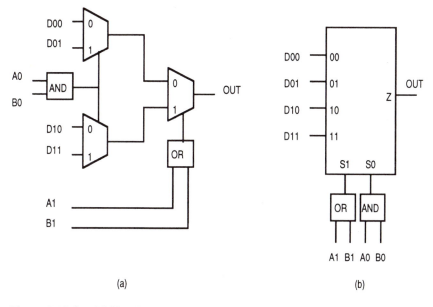

(a)   (b)

Figure 1.12 Act-2 "C" cell

The "S" cell, shown in Figure 1.13, consists of a front end equivalent to "C" cell followed by sequential block built around two latches. The sequential block can be used as a rising- or falling-edge D flip-flop or a transparent-high or transparent-low latch, by tying the C1 and C2 inputs to a clock signal, logical zero or logical one in various combinations. For example, tying C1 to 0 and clocking C2 implements a rising-edge D flip-flop. Toggle or enabled flip-flops can be built using combinational front end in addition to the D flip-flop. JK or SR flip-flops can be configured from one or more "C" or "S" cells using external feedback connections. A chip with an equal mixture of "C" and "S" cells provides sufficient flip-flops for most designs plus extra flexibility in placement. Over a range of designs, the Act-2 mixture provides about 40-100% greater logic capacity per cell than the Act-1 cell.

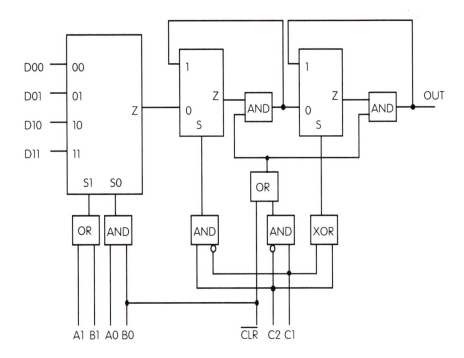

Figure 1.13 Actel-2 "S" cell

The logic cell in the FPLD from QuickLogic is similar to the Actel logic cell in that it uses a 4-to-1 multiplexer. Each input to the multiplexer is fed by an AND gate, as shown in Figure 1.14. Alternating inputs to the AND gates are inverted allowing input signals to be passed in true or complement form, therefore eliminating the need to use extra logic cells to perform simple inversions.

Multiplexer-based logic cells provide a large degree of functionality for a relatively small number of transistors. However, this is achieved at the expense of a large number of inputs placing high demands on the routing resources. They are best suited to FPLDs that use small size programmable switches such as antifuses.

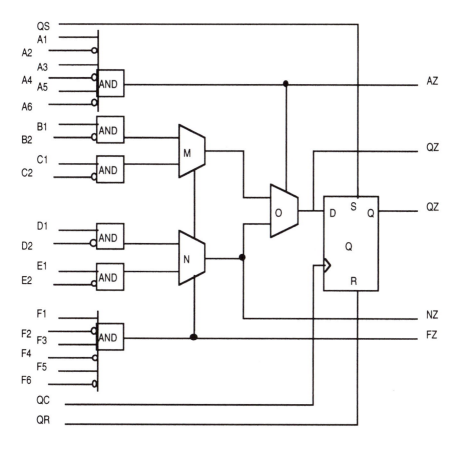

Figure 1.14 The QuickLogic logic cell

Xilinx logic cells are based on the use of SRAM as a look-up table. The truth table for a K-input logic function is stored in a 2^K x 1 SRAM as it is illustrated in Figure 1.15.

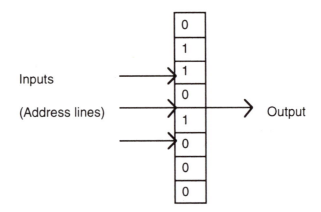

Figure 1.15 Look-up table

The address lines of the SRAM function as inputs and the output (data) line of the SRAM provides the value of the logic function. The major advantage of K-input look-up table is that it can implement any function of K inputs. The disadvantage is that it becomes unacceptably large for more than five inputs since the number of memory cells needed for a K-input look-up table is 2^K. Since many of the logic functions are not commonly used, a large look-up table will be largely underutilized.

The Xilinx 3000 series logic cell contains a five input one output look-up table. This block can be configured as two four-input LUTs if the total number of distinct inputs is not greater than five. The logic cell also contains sequential logic (two D flip-flops) and several multiplexers that connect combinational inputs and outputs to the flip-flops or outputs.

The Xilinx 4000 series logic cell contains two four input look-up tables feeding into a three input LUT. All of the inputs are distinct and available external to the logic cell. The other difference from the 3000 series cell is the use of two nonprogrammable connections from the two four input LUTs to the three input LUT. These connections are much faster since no programmable switches are used in series.

A detailed explanation of Xilinx 3000 and 4000 series logic cells is given in Chapter 2, since they represent two of the most popular and widely used FPGAs.

Other popular families of FPLDs with the coarse-grain logic cells are Altera's EPLDs and CPLDs. The architecture of Altera 5000 and 7000 series EPLDs has evolved from a PLA-based architecture with logic cells consisting of wide fan-in (20 to over 100 inputs) AND gates feeding into an OR gate with three to eight inputs. They employ a floating gate transistor based programmable switch that enables an input wire to be connected to an input to the gate as shown in Figure 1.16. The three product terms are then OR-ed together and can be programmable inverted by an XOR gate which can also be used to produce other arithmetic functions. Each signal is provided in both truth and complement form with two separate wires. The programmable inversion significantly increases the functional capability of the block.

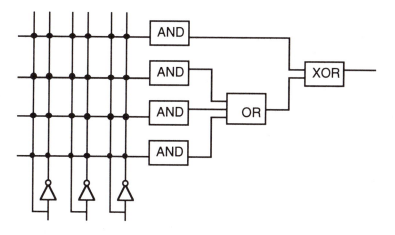

Figure 1.16 The Altera 5000 series logic block

The advantage of this type of block is that the wide AND gate can be used to form logic functions with few levels of logic cells reducing the need for programmable interconnect resources. However, it is difficult to make efficient use of the inputs to all of the gates. This loss is compensated by the high packing density of the wired AND gates. Some shortcomings of the 5000 series devices are overcome in the 7000 series, most notably it provides two more product terms and has more flexibility because neighboring blocks can "borrow" product terms from each other.

The Altera Flex 8000 and 10K series CPLDs are the SRAM based devices providing low stand-by power and in-circuit reconfigurability. A logic cell contains 4-input LUT that provides combinational logic capability and a programmable register that offers sequential logic capability. High system performance is provided by a fast, continuous network of routing resources. The detailed description of both major Altera's series of CPLDs is given in Chapter 2.

Most of the logic cells described above include some form of sequential logic. The Xilinx devices have two D flip-flops, while the Altera devices have one D flip-flop per logic cell. Some devices such as Act-1 do not explicitly include sequential logic, forming it using programmable routing and combinational logic cells.

1.5 Routing Architecture

The routing architecture of an FPLD determines a way in which the programmable switches and wiring segments are positioned to allow the programmable interconnection of logic cells. A routing architecture for an FPLD must meet two criteria: routability and speed. Routability refers to the capability of an FPLD to accommodate all the nets of a typical application, despite the fact that the wiring segments must be defined at the time the blank FPLD is made. Only switches connecting wiring segments can be programmed (customized) for a specific application, not the numbers, lengths or locations of the wiring segments themselves. The goal is to provide a sufficient number of wiring segments while not wasting chip area. It is also important that the routing of an application can be determined by an automated algorithm with minimal intervention.

Propagation delay through the routing is a major factor in FPLD performance. After routing an FPLD, the exact segments and switches used to establish the net are known and the delay from the driving output to each input can be computed. Any programmable switch (EPROM, pass-transistor, or antifuse) has a significant resistance and capacitance. Each time a signal passes through a programmable switch, another RC stage is added to the propagation delay. For a fixed R and C, the propagation delay mounts quadratically with the number of series RC stages. The use of a low resistance switch, such as antifuse, keeps the delay low and its distribution tight. Of equal significance is optimization of the routing architecture. Routing architectures of some commercial FPLD families are presented in this section.

In order to present commercial routing architectures, we will use the routing architecture model shown in Figure 1.17. First, a few definitions are introduced in order to form a unified viewpoint when considering routing architectures.

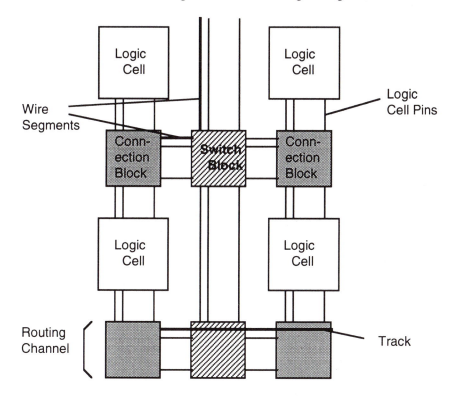

Figure 1.17 General FPLD routing architecture model

A wire segment is a wire unbroken by programmable switches. One or more switches may attach to a wire segment. Typically, one switch is attached to the each end of a wire segment. A track is a sequence of one or more wire segments in a line. A routing channel is a group of parallel tracks.

As shown in Figure 1.17, the model contains two basic structures. The first is a connection block which appears in all architectures. The connection block provides connectivity from the inputs and outputs of a logic block to the wire segments in the channels and can be both vertical or horizontal. The second structure is the switch block which provides connectivity between the horizontal as well as vertical wire segments. The switch block in Figure 1.17 provides connectivity among wire segments on all four of its sides.

Trade-offs in routing architectures are illustrated in Figure 1.18. Figure 1.18(a) represents a set of nets routed in a conventional channel. Freedom to configure the wiring of an MPLD allows us to customize the lengths of horizontal wires.

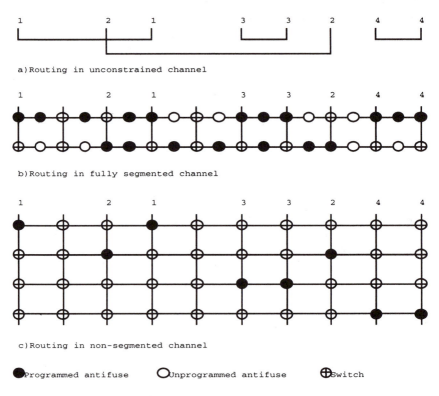

a) Routing in unconstrained channel

b) Routing in fully segmented channel

c) Routing in non-segmented channel

● Programmed antifuse   ○ Unprogrammed antifuse   ⊕ Switch

Figure 1.18 Types of routing architecture

In order to have complete freedom of routing, a switch is required at every cross point. More switches are required between two cross points along a track to allow the track to be subdivided into segments of arbitrary length, as shown in Figure 1.18(b). In FPLDs, each signal enters or leaves the channel on its own vertical segment.

An alternative is to provide continuous tracks in sufficient number to accommodate all nets, as shown in Figure 1.18(c). This approach is used in many types of programmable logic arrays and in the interconnect portion of certain programmable devices. Advantages are that two RC stages are encountered and that the delay of each net is identical and predictable. However, full length tracks are used for all, even short nets. Furthermore, the area is excessive, growing quadratically with the number of nets. This is the reason to employ some intermediate approaches, usually based on segmentation of tracks into varying (appropriate) sizes. A well-designed segmented channel does not require many more tracks than would be needed in a conventional channel. Although surprising, this finding has been supported both experimentally and analytically.

In the Xilinx 3000 series FPGAs, the routing architecture connections are made from the logic cell to the channel through a connection block. Since the connection site is large, because of the SRAM programming technology, the Xilinx connection block typically connects each pin to only two or three out of five tracks passing a cell. Connection blocks connect on all four sides of the cell. The connections are implemented by pass transistors for the output pins and multiplexers for input pins. The use òf multiplexers reduces the number of SRAM cells needed per pin.

The switch block makes a connection between segments in intersecting horizontal and vertical channels. Each wire segment can connect to a subset of the wire segments on opposing sides of the switch block (typically to 5 or 6 out of 15 possible wire segments). This number is limited by the large size and capacitance of the SRAM programmable switches.

There are four types of wire segments provided in the Xilinx 3000 architecture and five types in the Xilinx 4000 architecture. The additional wire segment consists of so called double-length lines that essentially represent the wire segments of the double length that are connected to every second switch block. In the Xilinx 4000 devices the connectivity between the logic cell pins and tracks is much higher because each logic pin connects to almost all of the tracks. The detailed presentation of the Xilinx routing architectures is given in Chapter 2.

The routing architecture of the Altera 5000 and 7000 series EPLDs uses a two-level hierarchy. At the first level hierarchy, 16 or 32 of the logic cells are grouped into a Logic Array Block (LAB) providing a structure very similar to the traditional PLD. There are four types of tracks passing each LAB. In the connection block every such track can connect into every logic cell pin making routing very simple. Using fewer connection points results in better density and performance, but yields more complex routing. The internal LAB routing structure could be considered as segmented channel, where the segments are as long as possible. Since connections also perform wire ANDing, the transistors have two purposes.

Connections among different LABs are made using a global interconnect structure called a Programmable Interconnect Array (PIA). It connects outputs from each LAB to inputs of other LABs, and acts as one large switch block. There is full connectivity among the logic cell outputs and LAB inputs within a PIA. The advantage of this scheme is that it makes routing easy, but requires many switches adding more to the capacitive load than necessary. Another advantage is the delay through the PIA is the same regardless of which track is used. This further helps predict system performance. However, the circuits can be much slower than with segmented tracks.

A similar approach is found in the Altera 8000 series CPLDs. Connections among LABs are implemented using FastTrack Interconnect continuous channels that run the length of the device. A detailed presentation of both of Altera's interconnect and routing mechanisms is given in Chapter 2.

1.6 Design Process

The complexity of FPLDs has surpassed the point where manual design is desirable or feasible. The utility of an FPLD architecture becomes more and more dependent on automated logic and layout synthesis tools.

The design process with FPLDs is similar to other programmable logic design. Input can come from a schematic netlist, a hardware description language, or a logic synthesis system. After defining what has to be designed, the next step is design implementation. It consists of fitting the logic into the FPLD structures. This step is called "logic partitioning" by some FPGA manufacturers and "logic fitting" in reference to CPLDs.

After partitioning, the design software assigns the logic, now described in terms of functional units on the FPLD, to a particular physical locations on the device and chooses the routing paths. This is similar to placement and routing traditional gate arrays.

One of the main advantages of FPLDs is their short development cycle compared to full- or semi-custom integrated circuits. Circuit design consists of three main tasks:

- design definition

- design implementation

- design modification

From the designer's point of view the following is important :

- the design process evolves toward behavioral level specification and synthesis

- design freedom from details of mapping to specific chip architecture

- easy way to change or correct design

A variety of design tools are used to perform all or some of the above tasks. Chapter 3 is devoted to the high level design tools with an emphasis on those that enable behavioral level specification and synthesis, primarily high-level hardware description languages. Examples of designs using two of such languages, the Altera Hardware description Language (AHDL) and VHSIC Hardware description Language (VHDL), are given together with the introduction to these specification tools.

An application targeted to an FPLD can be designed on any one of a number of logic or ASIC design systems, including schematic capture and hardware description languages. To target an FPLD, the design is passed to FPLD specific implementation software. The interface between design entry and design implementation is a netlist that contains the desired nets, gates, and references to specific vendor provided macros. Manual and automatic tools can be used interchangeably or an implementation can be done fully automatically.

A combination of moderate density, reprogrammability and powerful prototyping tools to a hardware designer resembles a software-like iterative-implementation methodology. Figure 1.19 is presented to compare a typical ASIC and typical FPLD design cycle.

In a typical ASIC design cycle, the design is verified by simulation at each stage of refinement. Accurate simulators are slow. ASIC designers use the whole range of simulators in the speed/accuracy spectrum in an attempt to verify their design. Although simulation can be used in designing for FPLDs, simulation can be replaced with in-circuit verification by simulating the circuitry in real time with a prototype. The path from design to prototype is short allowing verification of the operation over a wide range of conditions at high speed and high accuracy.

A fast design-place-route-load loop is similar to the software edit-compile-run loop and provides similar benefits, a design can be verified by the trial and error method. A designer can also verify that a design works in a real system, not merely in a potentially erroneous simulation.

Design by prototype does not verify proper operation with worst case timing, but rather that a design works on the typical prototype part. To verify worst case timing, designers can check speed margins in actual voltage and temperature corners with a scope and logic analyzer, speeding up marginal signals. They also may use a software timing analyzer or simulator after debugging to verify worst case paths or simply use faster speed grade parts in production to ensure sufficient speed margins over the complete temperature and voltage range.

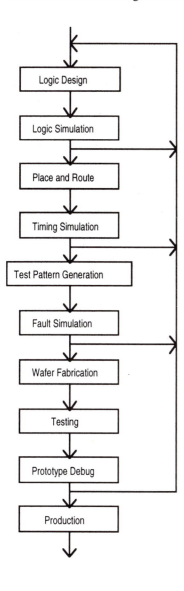

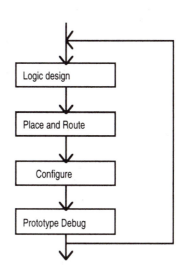

Design Cycle for FPLDs

Traditional ASIC Design Cycle

Figure 1.19 Comparing Design Cycles

As with software development, a reprogrammable FPLD removes the dividing line between prototyping and production. A working prototype may qualify as a production part if it meets performance and cost goals. Rather than redesign, a designer may choose to substitute a faster FPLD and use the same programming bitstream, or choose a smaller, cheaper FPLD (with more manual work to squeeze the design into a smaller device). A third choice is to substitute a mask-programmed version of the logic array for the field-programmable array. All three options are much simpler than a system redesign, which must be done for traditional MPLDs or ASICs.

The design process usually begins with the capture of the design. Most users enter their designs as schematics built of macros from a library. An alternative is to enter designs in terms of Boolean equations, state machine descriptions, or functional specifications. Different portions of a design can be described in different ways, compiled separately, and merged at some higher hierarchical level in the schematic.

Several guidelines are suggested for reliable design with FPLDs, mostly the same as those for users of MPLDs. The major goal is to make the circuit function properly independent of shifts in timing from one part to the next. Guidelines will be discussed in Chapter 3.

Rapid system prototyping is most effective when it becomes rapid product development. Reprogrammability allows a system designer another option, to modify the design in an FPLD by changing the programming bitstream after the design is in the hands of the customer. The bitstream can be stored in a dedicated (E)PROM or elsewhere in the system. In some existing systems, manufacturers send modified versions of hardware on a floppy disk or as a file sent over modem.

2 EXAMPLES OF FPLD FAMILIES

In this Chapter we will concentrate on a detailed description of the two major FPLD families from Altera and Xilinx. Their popularity comes from the high flexibility of individual devices, high circuit densities, reconfigurability, as well as the range of design tools. We will emphasize the most important features found in FPLDs and their use in complex digital system design and prototyping.

2.1 The Altera MAX 7000 Devices

As mentioned in Chapter 1, Altera has two different types of FPLDs: general purpose Erasable Programmable Logic Devices (EPLDs) based on floating gate programming technology used in the MAX 5000 and 7000 series, and SRAM based Flexible Logic Element matriX (FLEX) 8000 and 10K series, called Complex Programmable Logic Devices (CPLDs). All Altera devices use CMOS process technology which provides lower power dissipation and greater reliability than bipolar technology. Currently, Altera's devices are built on an advanced 0.5 and 0.35 µm technology. EPLDs from the 5000 and 7000 series are targeted for combinatorially intensive logic designs. In the following sections we will concentrate on the MAX 7000, the more advanced series. This family provides densities ranging from 150 to 5,000 usable logic gates and pin counts ranging from 44 to 208 pins. The FLEX 8000 family provides logic density from 2,500 to 24,000 usable gates and pin counts from 84 to 304 pins. Predictable interconnect delays combined with the high register counts, low standby power, and in-circuit reconfigurability of FLEX 8000 make these devices suitable for high density, register intensive designs. The FLEX 10K family provides logic densities of up to more than 100,000 gates.

2.1.1 Altera MAX 7000 devices general concepts

The MAX 7000 family of high density, high performance EPLDs provides dedicated input pins, user configurable I/O pins, programmable flip-flops, and clock options that ensure flexibility for integrating random logic functions. The MAX 7000 architecture supports emulation of standard TTL circuits and integration of SSI, MSI, and LSI logic functions. It also easily integrates multiple programmable logic devices from standard PALs, GALs, and FPLDs. MAX 7000 EPLDs use CMOS EEPROM cells as the programming technology to implement logic functions and contain from 32 to 256 logic cells, called macrocells, in groups of 16 macrocells, also called Logic Array Blocks (LABs). Each macrocell has a programmable-AND/fixed-OR array and a configurable register with independently programmable Clock, Clock Enable, Clear, and Preset functions.

Each EPLD contains an AND array that provides product terms, which are essentially n-input AND gates. EPLD schematics use a shorthand AND-array notation to represent several large AND gates with common inputs. An example of the same function in different representations is shown in Figure 2.1. In Figures 2.1(a), (b), and (c), a classic, sum-of-product and AND array notation are shown, respectively. A dot represents a connection between an input (vertical line) and one of the inputs to an n-input AND gate. If there is no connection to the AND gate, AND gate input is unused and floats to a logic 1.

The AND array circuit of Figure 2.1(c), with two 8-input AND gates, can produce any Boolean function of four variables (provided that only two product terms or simply p-terms are required) when expressed in sum-of-products form. The outputs of the product terms are tied to the inputs of an OR gate to compute the sum.

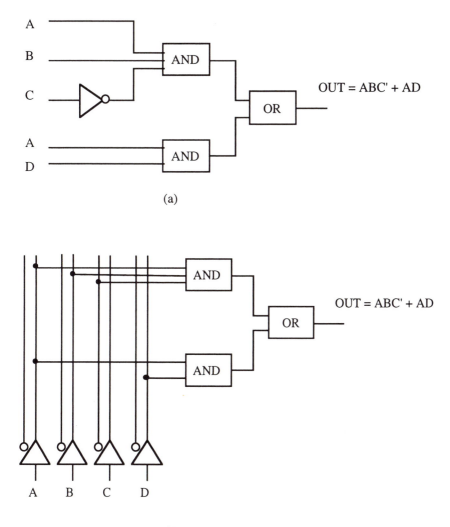

(a)

(b)

Figure 2.1 Different representations of logic function

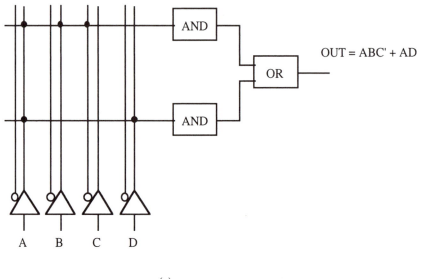

(c)

Figure 2.1 Different representations of logic function (cont.)

Product terms can also be used to generate complex control signals for use with programmable registers (Clock, Clock Enable, Clear, and Preset) or Output Enable signals for the I/O pins. These signals are called array control signals.

As discussed in Chapter 1, the Altera EPLDs support programmable inversion allowing software to generate inversions, wherever necessary, without wasting macrocells for simple functions. Software also automatically applies De Morgan's inversion and other logic synthesis techniques to optimize the use of available resources.

In the remaining sections of this Chapter we will present the functional units of the Altera EPLD in enough detail to understand their operation and application potential.

2.1.2 Macrocell

The fundamental building block of an Altera EPLD is the macrocell. A MAX 7000 macrocell can be individually configured for both combinational and sequential operation. Each macrocell consists of three parts:

- A logic array that implements combinational logic functions

- A product term select matrix that selects product terms which take part in implementation of logic function

- A programmable register that provides D, T, JK, or SR options that can be bypassed

One typical macrocell architecture of MAX 7000 series is shown in Figure 2.2.

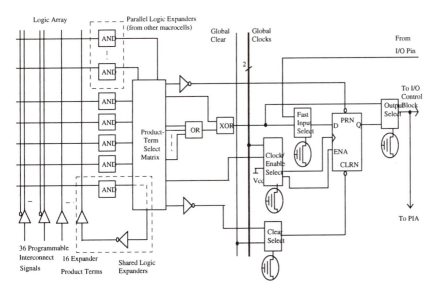

Figure 2.2 Macrocell architecture

A logic array consists of a programmable AND/ fixed OR array, known as PLA. Inputs to the AND array come from the true and complement of the dedicated input and clock pins from macrocell paths and I/O feedback paths. A typical logic array contains 5 p-terms that are distributed among the combinational and sequential resources. Connections are opened during the programming process. Any p-term may be connected to the true and complement of any array input signal. The p-term select matrix allocates these p-terms for use as either primary logic inputs (to the OR and XOR gates) to implement logic functions or as secondary inputs to the macrocell's register Clear, Preset, Clock, and Clock Enable functions. One p-term per macrocell can be inverted and fed back into the logic array. This "shareable" p-term can be connected to any p-term within the LAB.

Each macrocell flip-flop can be programmed to emulate D, T, JK, or SR operations with a programmable Clock control. If necessary, a flip-flop can be bypassed for combinational (non-registered) operation and can be clocked in three different modes.

- The first is by a global clock signal. This mode results with the fastest Clock to output performance.

- The second is by a global Clock signal and enabled by an active high Clock Enable. This mode provides an Enable on each flip-flop while still resulting in the fast Clock to output performance of the global Clock.

- Finally, the third is by an array Clock implemented with a p-term. In this mode, the flip-flop can be clocked by signals from buried macrocells or I/O pins.

Each register also supports asynchronous Preset and Clear functions by the p-terms selected by the p-term select matrix. Although the signals are active high, active low control can be obtained by inverting signals within the logic array. In addition, the Clear function can be driven by the active low, dedicated global Clear pin.

The flip-flops in macrocells also have a direct input path from the I/O pin, which bypasses PIA and combinational logic. This input path allows the flip-flop to be used as an input register with a fast input set up time (3 ns).

The more complex logic functions, those requiring more than five p-terms, can be implemented using shareable and parallel expander p-terms instead of additional macrocells. These expanders provide outputs directly to any macrocell in the same LAB.

Each LAB has up to 16 shareable expanders that can be viewed as a pool of uncommitted single p-terms (one from each macrocell) with inverted outputs that feed back into the logic array. Each shareable expander can be used and shared by any macrocell in the LAB to build complex logic functions.

Parallel expanders are unused p-terms from macrocells that can be allocated to a neighboring macrocells to implement fast, complex logic functions. Parallel expanders allow up to 20 p-terms to directly feed the macrocell OR logic five p-terms are provided by the macrocell itself and 15 parallel expanders are provided by neighboring macrocells in the LAB.

When both the true and complement of any signal are connected intact, a logic low 0 results on the output of the p-term. If both the true and complement are open, a logical "don't care" results for that input. If all inputs for the p-term are programmed opened, a logic high (1) results on the output of the p-term.

Several p-terms are input to a fixed OR whose output connects to an exclusive OR (XOR) gate. The second input to the XOR gate is controlled by a programmable resource (usually a p-term) that allows the logic array output to be inverted. In this way active low or active high logic can be implemented, as well as the number of p-terms can be reduced (by applying De Morgan's inversion).

2.1.3 I/O Control Block

The EPLD I/O control block contains a tri state buffer controlled by one of the global Output Enable signals or directly connected to GND or Vcc, as shown in Figure 2.3. When the tri state buffer control is connected to GND, the output is in high impedance and the I/O pin can be used as a dedicated input. When the tri-state buffer control is connected to Vcc, the output is enabled.

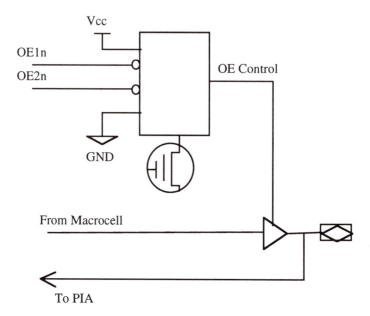

Figure 2.3 I/O control block

I/O pins may be configured as dedicated outputs, bi-directional lines, or as additional dedicated inputs. Most EPLDs have dual feedback, with macrocell feedback being decoupled from the I/O pin feedback.

In the high end devices from the MAX 7000 family the I/O control block has six global Output Enable signals that are driven by the true or complement of two Output Enable signals (a subset of the I/O pins) or a subset of the I/O macrocells. This is shown in Figure 2.4. Macrocell and pin feedbacks are independent. When an I/O pin is configured as an input, the associated macrocell can be used for buried logic.

Additional features are found in the MAX 7000 series. Each macrocell can be programmed for either high speed or low power operation. The output buffer for each I/O pin has an adjustable output slew rate that can be configured for low noise or high speed operation. The fast slew rate should be used for speed critical outputs in systems that are adequately protected against noise.

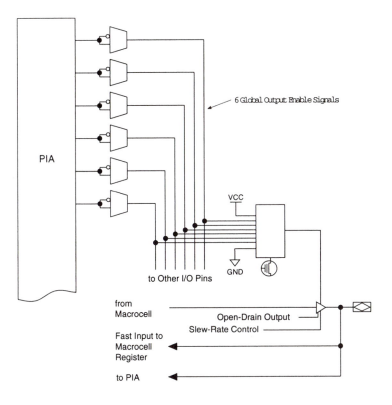

Figure 2.4 I/O control block in high end MAX 7000 devices

2.1.4 Logic Array Blocks

Programmable logic in EPLDs is organized into Logic Array Blocks (LABs). Each LAB contains a macrocell array, an expander product term array, and an I/O control block. The number of macrocells and expanders varies with each device. The general structure of the LAB is presented in Figure 2.5. Each LAB is accessible through Programmable Interconnect Array (PIA) lines and input lines. Macrocells are the primary resource for logic implementation, but expanders can be used to supplement the capabilities of any macrocell. The outputs of a macrocell feed the decoupled I/O block, which consists of a group of programmable 3-state buffers and I/O pins. Macrocells that drive an output pin may use the Output Enable p-term to control the active high 3-state buffer in the I/O control block. This allows complete and exact emulation of 7400 series TTL family.

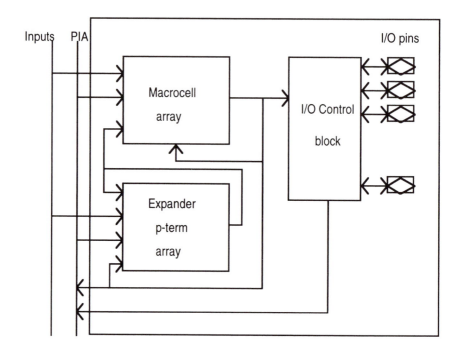

Figure 2.5 Logic Array Block architecture

Each LAB has two clocking modes: asynchronous and synchronous. During asynchronous clocking, each flip-flop is clocked by a p-term allowing that any input or internal logic to be used as a clock. Moreover, each flip-flop can be configured for positive or negative edge triggered operation.

Synchronous clocking is provided by a dedicated system clock (CLK). Since each LAB has one synchronous clock, all flip-flop clocks within it are positive edge triggered from the CLK pin.

Altera EPLDs have an expandable, modular architecture allowing several hundreds to tens of thousands of gates in one package. They are based on a logic matrix architecture consisting of a matrix of LABs connected with a PIA, shown in Figure 2.6. The PIA provides a connection path with a small fixed delay between all internal signal sources and logic destinations.

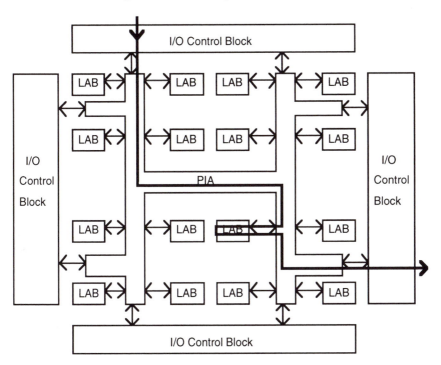

Figure 2.6 Altera EPLD entire block diagram

2.1.5 Programmable Interconnect Array

Logic is routed between LABs on the Programmable Interconnect Array (PIA). This global bus is programmable and enables connection of any signal source to any destination on the device. All dedicated inputs, I/O pins, and macrocell outputs feed the PIA, which makes them available throughout the entire device. An EEPROM cell controls one input of a 2-input AND gate which selects a PIA signal to drive into the LAB, as shown in Figure 2.7. Only signals required by each LAB are actually routed into the LAB.

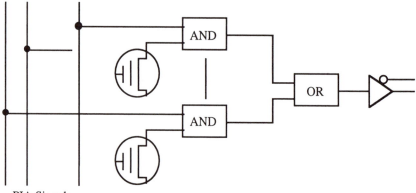

PIA Signals

Figure 2.7 PIA routing

While routing delays in channel based routing schemes in MPGAs and FPGAs are cumulative, variable, and path dependent, the MAX PIA has a fixed delay. Therefore, it eliminates skew between signals and makes timing performance easy to predict. MAX 7000 devices have fixed internal delays allowing the user to determine the worst case timing for any design.

2.1.6 Programming

Programming of MAX 7000 devices consists of configuring EEPROM transistors as required by design. The normal programming procedure consists of the following steps:

1. The programming pin (V_{pp}) is raised to the super high input level (usually 12.5V).

2. Row and column address are placed on the designated address lines (pins).

3. Programming data is placed on the designated data lines (pins).

4. The programming algorithm is executed with a sequence of 100 microsecond programming pulses separated by program verify cycles.

5. Overprogram or margin pulses may be applied to double ensure EPLD programming.

The programming operation is typically performed eight bits at a time on specialized hardware. The security bit can be set to ensure EPLD design security.

Some of the devices from the MAX 7000 family have special features such as 3.3 V operation or power management. The 3.3 V operation offers power savings of 30% to 50% over 5.0 V operation. The power saving features include a programmable power saving mode and power down mode. Power down mode allows the device to consume near zero power (typically 50 μA). This mode of operation is controlled externally by the dedicated power down pin. When this signal is asserted, the power down sequence latches all input pins, internal logic, and output pins preserving their present state.

2.2 The Altera FLEX 8000 and 10K Devices

Altera's Flexible Logic Element Matrix (FLEX) programmable logic combines the high register counts of CPLDs and the fast predictable interconnects of EPLDs. It is SRAM based providing low stand by power and in circuit reconfigurability. Logic is implemented with 4-input look-up tables (LUTs) and programmable registers. High performance is provided by a fast, continuous network of routing resources. FLEX 8000 devices are configured at system power up, with data stored in a serial configuration EPROM device or provided by a system controller. Configuration data can also be stored in an industry standard EPROM or downloaded from system RAM. Since reconfiguration requires less than 100 ms, real time changes can be made during system operation.

The FLEX architecture incorporates a large matrix of compact logic cells called logic elements (LEs). Each LE contains a 4-input LUT that provides combinatorial logic capability each also contains a programmable register that offers sequential logic capability. LEs are grouped into sets of eight to create Logic Array Blocks (LABs). Each LAB is an independent structure with common inputs, interconnections, and control signals. LABs are arranged into rows and columns. The I/O pins are supported by I/O elements (IOEs) located at the ends of rows and columns. Each IOE contains a bi-directional I/O buffer and a flip-flop that can be used as either an input or output register. Signal interconnections within FLEX 8000 devices are provided by FastTrack Interconnect continuous channels that run the entire length and width of the device. The architecture of FLEX 8000 device is illustrated in Figure 2.8.

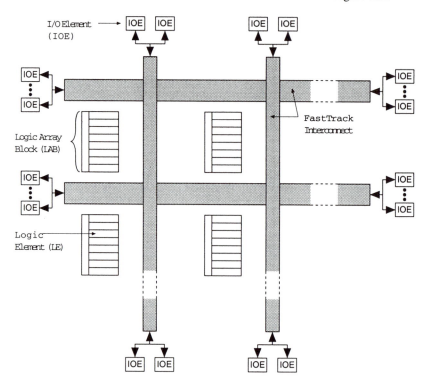

Figure 2.8 FLEX 8000 device architecture

2.2.1 Logic Element

A logic element (LE) is the basic logic unit in the FLEX 8000 architecture. Each LE contains a 4-input LUT, a programmable flip-flop, a carry chain, and a cascade chain as shown in Figure 2.9.

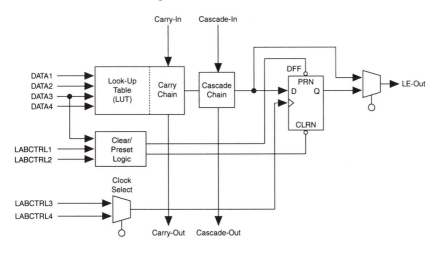

Figure 2.9 Logic element

The LUT quickly computes any Boolean function of four input variables. The programmable flip-flop can be configured for D, T, JK, or SR operation. The Clock, Clear, and Preset control signals can be driven by dedicated input pins, general purpose I/O pins, or any internal logic. For combinational logic, the flip-flop is bypassed and the output of the LUT goes directly to the output of the LE.

Two dedicated high speed paths are provided in the FLEX 8000 architecture; the carry chain and cascade chain both connect adjacent LEs without using general purpose interconnect paths. The carry chain supports high speed adders and counters. The cascade chain implements wide input functions with minimal delay. Carry and cascade chains connect all LEs in a LAB and all LABs of the same row.

The carry chain provides a very fast (less than 1 ns) carry forward function between LEs. The carry-in signal from a lower order bit moves towards the higher order bit by way of the carry chain and also feeds both the LUT and a portion of the carry chain of the next LE. This feature allows implementation of high speed counters and adders of practically arbitrary width. A 4-bit parallel full adder can be implemented in 4+1=5 LEs by using the carry chain as shown in Figure 2.10. The LEs look-up table is divided into two portions. The first portion generates the sum of two bits using input signals and the carry-in signal. The other generates the carry-out signal, which is routed directly to the carry-in input of the next higher order bit. The final carry-out signal is routed to an additional LE, and can be used for any purpose.

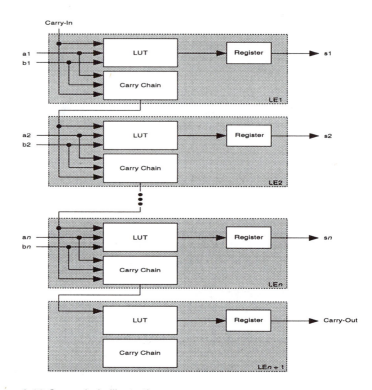

Figure 2.10 Carry chain illustration

With the cascade chain, the FLEX 8000 architecture can implement functions with a very wide fan-in. Adjacent LUTs can be used to compute portions of the function in parallel, while the cascade chain serially connects the intermediate values. The cascade chain can use a logical AND or logical OR to connect the outputs of adjacent LEs. Each additional LE provides four more inputs to the effective width of a function adding a delay of approximately 1 ns per LE.

Figure 2.11 illustrates how the cascade function can connect adjacent LEs to form functions with wide fan-in.

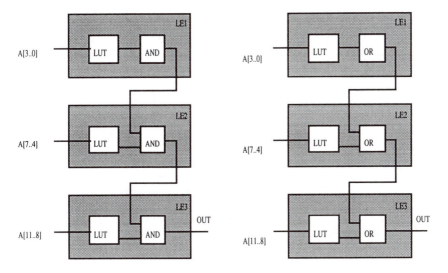

Figure 2.11 Cascade chain illustration

The LE can operate in the four different modes (shown in Figure 2.12). In each mode, seven of the ten available inputs to the LE - the four data inputs from the LAB, local interconnect, the feedback from the programmable register, and the carry-in from the previous LE are directed to different destinations to implement the desired logic function. The remaining inputs provide control for the register.

The normal mode is suitable for general logic applications and wide decode functions that can take advantage of a cascade chain.

Normal Mode

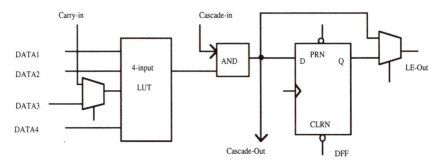

Arithmetic Mode

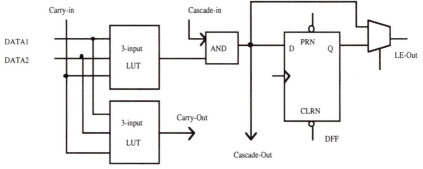

Up/Down Counter Mode

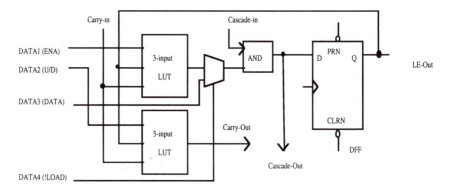

Clearable Counter Mode

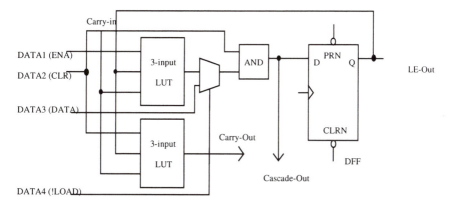

Figure 2.12 LE operating modes

The arithmetic mode offers two 3-input LUTs that are ideal for implementing adders, accumulators, and comparators. One LUT provides a 3-bit Boolean function, and the other generates a carry bit. The arithmetic mode also supports a cascade chain.

The Up/Down counter mode offers counter enable, synchronous up/down control, and data loading options. Two 3-input LUTs are used: one generates the counter data, the other generates the fast carry bit. A 2-to-1 multiplexer provides synchronous loading. Data can also be loaded asynchronously with the Clear and Preset register control signals.

The clearable counter mode is similar to the Up/Down counter mode, but supports a synchronous Clear instead of the up/down control. The Clear function is substituted for Cascade-in signal in Up/Down Counter mode. Two 3-input LUTs are used: one generates the counter data, the other generates the fast carry bit.

The Logic controlling a register's Clear and Preset functions is controlled by the DATA3, LABCTRL1, and LABCTRL2 inputs to LE, as shown in Figure 2.13. Default values for the Clear and Preset signals, if unused, are logic highs.

(a) Clear Logic

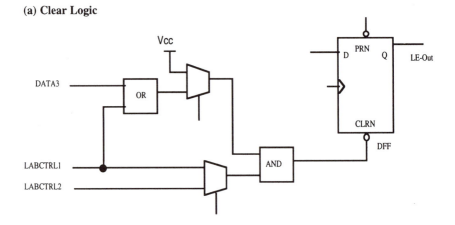

(b) Preset Logic

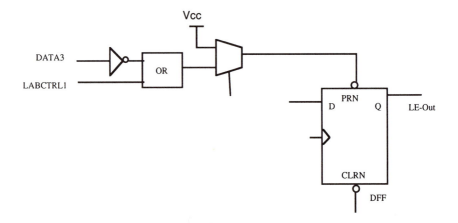

Figure 2.13 LE Clear and Preset Logic

If the flip-flop is cleared by only one of two LABCTRL signals, the DATA3 input is not required and can be used for one of the logic element operating modes.

2.2.2 Logic Array Block

A Logic Array Block (LAB) consists of eight LEs, their associated carry chains, cascade chains, LAB control signals, and the LAB local interconnect. The LAB structure is illustrated in Figure 2.14. Each LAB provides four control signals that can be used in all eight LEs. Two of these signals can be used as Clocks and the other two for Clear/Preset control. The LAB control signals can be driven directly from a dedicated I/O pin or any internal signal by way of the LAB local interconnect. The dedicated inputs are typically used for the global Clock, Clear, and Preset because they provide synchronous control with very low skew across the device. If logic is required on a control signal, it can be generated in one or more LEs in any LAB and driven into the local interconnect of the target LAB. Programmable inversion is available for all four LAB control signals.

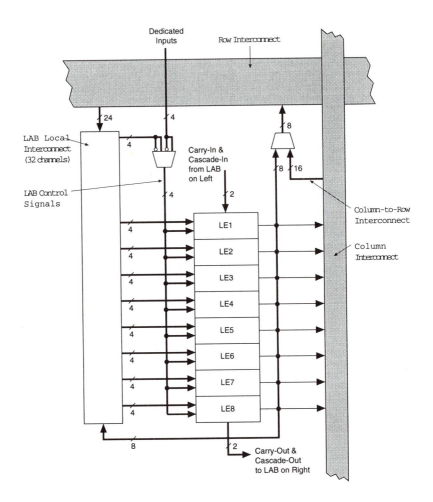

Figure 2.14 LAB Internal Architecture

2.2.3 FastTrack Interconnect

Connections between LEs and device I/O pins are provided by the FastTrack Interconnect mechanism represented by a series of continuous horizontal and vertical routing channels that traverse the entire device. The LABs within the device are arranged into a matrix of columns and rows. Each row has a dedicated interconnect that routes signals into and out of the LABs in the row. The row interconnect can then drive I/O pins or feed other LABs in the device. Figure 2.15 shows how an LE drives the row and column interconnect.

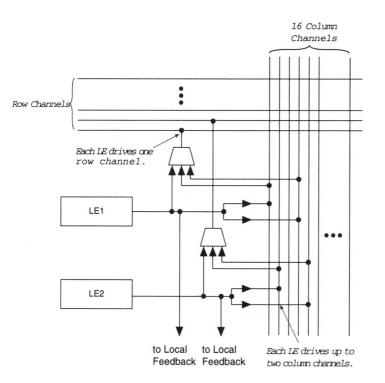

Figure 2.15 LAB Connections to Row and Column Interconnect

Each LE in a LAB can drive up to two separate column interconnect channels. Therefore, all 16 available column channels can be driven by a LAB. The column channels run vertically across the entire device and LABs in different rows share access to them by way of partially populated multiplexers. A row interconnect channel can be fed by the output of the LE or by two column channels. These three signals feed a multiplexer that connects to a specific row channel. Each LE is connected to one 3-to-1 multiplexer. In a LAB, the multiplexers provide all 16 column channels with access to the row channels.

Each column of LABs has a dedicated column interconnect that routes signals out of the LABs in that column. The column interconnect can drive I/O pins or feed into the row interconnect to route the signals to other LABs in the device. A signal from the column interconnect, which can be either the output from an LE or an input from an I/O pin, must transfer to the row interconnect before it can enter a LAB. Figure 2.16 shows the interconnection of four adjacent LABs with row, column, and local interconnects, as well as associated cascade and carry chains.

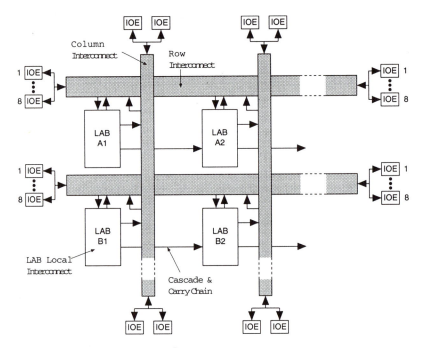

Figure 2.16 Device Interconnect Resources

The Interconnection between row interconnect channels and IOEs is illustrated in Figure 2.17. An input signal from an IOE can drive two row channels. When an IOE is used as an output, the signal is driven by an n-to-1 multiplexer that selects the row channels. The size of the multiplexer depends on the number of columns in the device. Eight IOEs are connected to each side of the row channels.

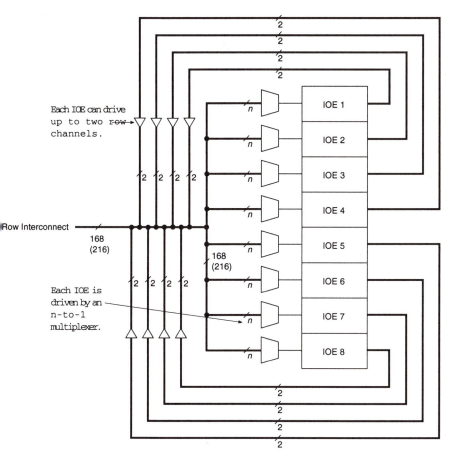

Figure 2.17 Row to IOE Connection

On the top and bottom of the column channels are two IOEs, as shown in Figure 2.18. When an IOE is used as an input, it can drive up to 2 column channels. The output signal to an IOE can choose from 8 column channels through an 8-to-1 multiplexer.

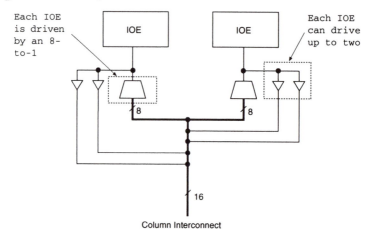

Figure 2.18 Column to IOE Connection

2.2.4 Dedicated I/O Pins

In addition to general purpose I/O pins, four dedicated input pins provide low skew and device wide signal distribution. Typically, they are used for global Clock, Clear, and Preset control signals. These signals are available for all LABs and IOEs in the device. The dedicated inputs can be used as general purpose data inputs for nets with large fan-outs because they feed the local interconnect.

2.2.5 Input/Output Element

Input/Output Element (IOE) architecture is presented in Figure 2.19. IOEs are located at the ends of the row and column interconnect channels. I/O pins can be used as input, output, or bidirectional pins. Each I/O pin has a register that can be used either as an input or output register in operations requiring high performance (fast set up time or fast Clock to output time). The output buffer in each IOE has an adjustable slew rate.

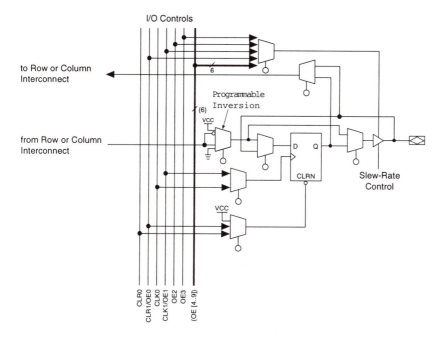

Figure 2.19 IOE Architecture

A fast slew rate should be used for speed critical outputs in systems protected against noise. Clock, Clear, and Output Enable controls for the IOE are provided by a network of I/O control signals. These signals are supplied by either the dedicated input pins or internal logic. All control signal sources are buffered onto high speed drivers that drive the signals around the periphery of the device. This "peripheral bus" can be configured to provide up to four Output Enable signals and up to two Clock or Clear signals.

The signals for the peripheral bus are generated by any of the four dedicated inputs or signals on the row interconnect channels, as shown in Figure 2.20.

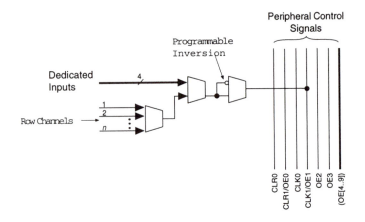

Figure 2.20 Peripheral Bus

The number of row channels used depends on the number of columns in the device. The six peripheral control signals can be accessed by every I/O element.

2.2.6 Configuring FLEX 8000 Devices

The FLEX 8000 family supports several configuration schemes for loading the design into a chip on the circuit board. The FLEX 8000 architecture uses SRAM cells to store configuration data for the device. These SRAM cells must be loaded each time the circuit powers up and begins operation. The process of physically loading the SRAM with programming data is called configuration. After configuration, the FLEX 8000 device resets its registers, enables I/O pins, and begins operation as a logic device. This reset operation is called initialization. Together, the configuration and initialization processes are called the command mode. Normal in circuit device operation is called the user mode.

The entire command model requires less than 100 ms and can be used to dynamically reconfigure the device even during system operation. Device configuration can occur either automatically at system power up or under control of external logic. The configuration data can be loaded into FLEX 8000 device with one of six configuration schemes, which is chosen on the basis of the target application.

There are two basic types of configuration schemes: active, and passive. In an active configuration scheme, the device controls the entire configuration process and generates the synchronization and control signals necessary to configure and initialize itself from external memory. In a passive configuration scheme, the device is incorporated into a system with an intelligent host that controls the configuration process. The host selects either a serial or parallel data source and the data is transferred to the device on a common data bus. The best configuration scheme depends primarily on the particular application and on factors such as the need to reconfigure in real time, the need to periodically install new configuration data, as well as other factors.

Generally, an active configuration scheme provides faster time to market because it requires no external intelligence. The device is typically configured at system power up, and reconfigured automatically if the device senses power failure. A passive configuration scheme is generally more suitable for fast prototyping and development (for example from development Max+PLUS II software) or in applications requiring real-time device reconfiguration. Reconfigurability allows reuse of logic resources instead of designing redundant or duplicate circuitry in a system. Short descriptions of several configuration schemes are presented in the following sections.

Active Serial Configuration

This scheme, with a typical circuit shown in Figure 2.21, uses Altera's serial configuration EPROM as a data source for FLEX 8000 devices. The nCONFIG pin is connected to Vcc, so the device automatically configures itself at system power up. Immediately after power up, the device pulls the nSTATUS pin low and releases it within 100 ms. The DCLK signal clocks serial data bits from the configuration EPROM. When the configuration is completed, the CONF_DONE signal is released causing the nCS to activate and bring the configuration EPROM data output into a high impedance state. After CONF_DONE goes high, the FLEX 8000 completes the initialization process and enters user mode. In the circuit shown in Figure 2.21, the nCONFIG signal is tied up to the Output Enable (OE) input of the configuration EPROM. External circuitry is necessary to monitor nSTATUS of the FLEX device in order to undertake appropriate action if configuration fails.

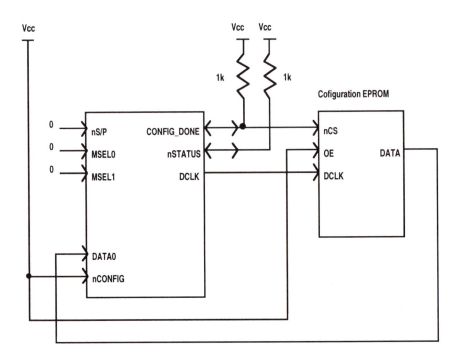

Figure 2.21 Active Serial Device Configuration

Active Parallel Up (APU) and Active Parallel Down (APD) Configuration

In Active Parallel Up and Active Parallel Down configuration schemes, the FLEX 8000 device generates sequential addresses that drive the address inputs to an external EPROM. The EPROM then returns the appropriate byte of data on the data lines DATA[7..0]. Sequential addresses are generated until the device has been completely loaded. The CONF_DONE pin is then released and pulled high externally indicating that configuration has been completed. The counting sequence is ascending (00000H to 3FFFFH) for APU or descending (3FFFFH to 00000H) for APD configuration. A typical circuit for parallel configuration is shown in Figure 2.22.

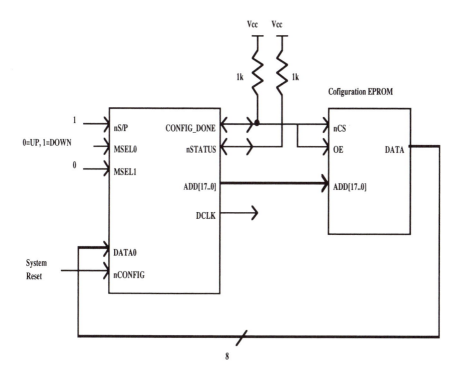

Figure 2.22 APU and APD Configuration with a 256 Kbyte EPROM

On each pulse of the RDCLK signal (generated by dividing DCLK by eight), the device latches an 8-bit value into a serial data stream. A new address is presented on the ADD[17..0] lines a short time after a rising edge on RDCLK. External parallel EPROM must present valid data before the subsequent rising edge of RDCLK, which is used to latch data based on address generated by the previous clock cycle.

Both active parallel configuration schemes can generate addresses in either an ascending or descending order. Counting up is appropriate if the configuration data is stored at the beginning of an EPROM or at some known offset in an EPROM larger of 256 Kbytes. Counting down is appropriate if the low addresses are not available, for example if they are used by the CPU for some other purpose.

Passive Parallel Synchronous Configuration

In this scheme the FLEX 8000 device is tied to an intelligent host. The DCLK, CONF_DONE, nCONFIG, and nSTATUS signals are connected to a port on the host, and the data can be driven directly onto a common data bus between the host and the FLEX 8000 device. New byte of data is latched on every eighth rising edge of DCLK signal, and serialized on every eight falling edge of this signal, until the device is completely configured.

A typical circuit for passive serial configuration is shown in Figure 2.23. The CPU generates a byte of configuration data. Data is usually supplied from a microcomputer 8-bit port. Dedicated data register can be implemented with an octal latch. The CPU generates clock cycles and data; eight DCLK cycles are required to latch and serialize each 8-bit data word and a new data word must be present at the DATA[7..0] inputs upon every eight DCLK cycles.

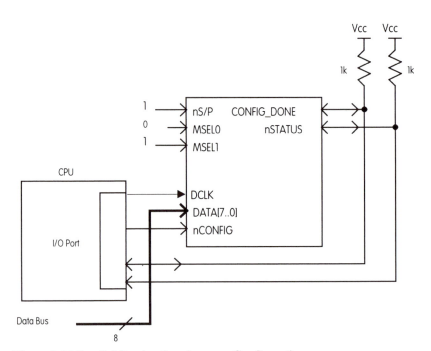

Figure 2.23 Parallel Passive Synchronous Configuration

Passive Parallel Asynchronous Configuration

In this configuration, a FLEX 8000 device can be used in parallel with the rest
of the board. The device accepts a parallel byte of input data, then serializes the
data with its internal synchronization clock. The device is selected with nCS
and CS chip select input pins. A typical circuit with a microcontroller as an
intelligent host is shown in Figure 2.24. Dedicated I/O ports are used to drive
all control signals and the data bus to the FLEX 8000 device. The CPU
performs handshaking with a device by sensing the RDYnBUSY signal to
establish when the device is ready to receive more data. The RDYnBUSY
signal falls immediately after the rising edge of the nWS signal that latches
data, indicating that the device is busy. On the eighth falling edge of DCLK,
RDYnBUSY returns to Vcc, indicating that another byte of data can be latched.

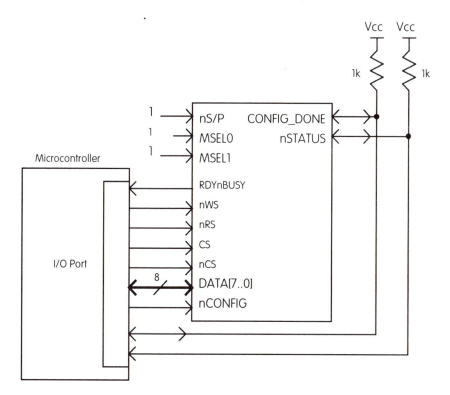

Figure 2.24 Passive Parallel Asynchronous Configuration

Passive Serial Configuration

The passive serial configuration scheme uses an external controller to configure the FLEX 8000 device with a serial bit stream. The FLEX device is treated as a slave and no handshaking is provided. Figure 2.25 shows how a bit wide passive configuration is implemented. Data bits are presented at the DATA0 input with the least significant bit of each byte of data presented first. The DCLK is strobed with a high pulse to latch the data. The serial data loading continues until the CONF_DONE goes high indicating that the device is fully configured. The data source can be any source that the host can address.

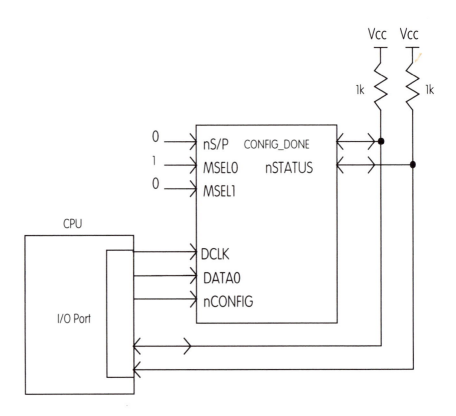

Figure 2.25 Bit Wide Passive Serial Configuration

2.2.7 Designing with FLEX 8000 Devices

In both EPLD and CPLD architectures, trade-offs are made to optimize designs for either speed or density. The FLEX 8000 architecture allows control of the speed/density trade-offs. In addition, Altera's Max+PLUS II software can automatically optimize all or part of a circuit for speed or density.

The Altera FLEX 8000 architecture is supported by design methods that offer a full spectrum of low to high level control over actual design implementation. If a fast design cycle is the primary goal, the design can be described with high level constructs in a hardware description language such as VHDL, Verilog, or AHDL.

If the designer wishes to optimize performance and density, the design can be described with primitive gates and registers ("gate level" design) using hardware description language or schematics. Family specific macrofunctions are also available.

Different logic options and synthesis styles can be used (set up) to optimize a design for a particular design family. Also different options can be used in portions of the design to improve the overall design. The following design guidelines yield maximum speed, reliability, and device resource utilization, while minimizing fitting problems.

1. Reserve Resources for Future Expansions. Because designs are modified and extended, we recommend leaving 20% of a device's logic cells and I/O pins unused.

2. Allow the Compiler to Select Pin & Logic Cell Assignment. Pin & logic cell assignments, if poorly or arbitrarily selected, can limit the Max+PLUS II compiler's ability to arrange signals efficiently, reducing the probability of a successful fit. We recommend the designer allow the compiler to choose all pin and logic cell locations automatically.

3. Balance Ripple Carry & Carry Look Ahead Usage. The dedicated carry chain in the FLEX 8000 architecture propagate a ripple carry for short and medium length counters and adders with minimum delay. Long carry chains, however, restrict the compiler's ability to fit a design because the LEs in the chain must be contiguous. On the other hand, look ahead

counters do not require the use of adjacent logic cells. This allows the compiler to arrange and permute the LEs to map the design into the device more efficiently.

4.   Use Global Clock & Clear Signals. Sequential logic is most reliable if it is fully synchronous, that is if every register in the design is clocked by the same global clock signal and reset by the same global clear signal. Four dedicated high speed, low skew global signals are available throughout the device, independent of FastTrack interconnect, for this purpose. Figure 2.13 shows the register control signals in the FLEX 8000 device. The Preset and Clear functions of the register can be functions of LABCTRL1, LABCTRL2, and DATA3. The asynchronous load and Preset are implemented within a single device. Figure 2.26 shows an asynchronous load with a Clear input signal. Since the Clear signal has priority over the load signal, it does not need to feed the Preset circuitry.

5.   Use One Hot Encoding of State Machines. One Hot Encoding (OHE) of states in state machines is a technique that uses one register per state and allows one state bit to be active at any time. Although this technique increases the number of registers, it also reduces the average fan-in to the state bits. In this way, the number of LEs required to implement the state decoding logic is minimized and OHE designs run faster and use less interconnect.

6.   Use Pipelining for Complex Combinatorial Logic. One of the major goals in circuit design is to maintain the clock speed at or above a certain frequency. This means that the longest delay path from the output of any register to the input(s) of the register(s) it feeds must be less than a certain value. If the delay path is too long, we recommend the pipelining of complex blocks of combinatorial logic by inserting flip-flops between them. This can increase device usage, but at the same time it lowers the propagation delay between registers and allows high system clock speeds. Pipelining is very effective especially with register intensive devices, such as FLEX 8000 devices.

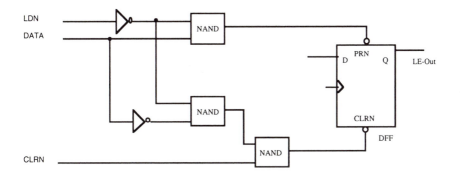

Figure 2.26 Asynchronous Load with Clear Input Signal

An asynchronous load without the Clear Input Signal is shown on Figure 2.27.

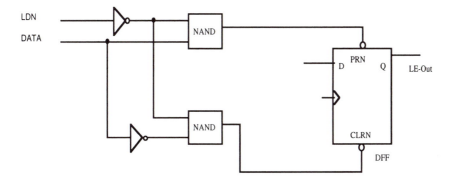

Figure 2.27 Asynchronous Load without a Clear Input Signal

An asynchronous Preset signal, which actually represents the load of a "1" into a register, is shown in Figure 2.28.

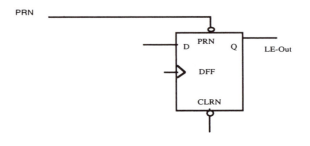

Figure 2.28 Asynchronous Preset

2.2.8 FLEX 10K Devices

The aim of this Section is to make a brief introduction to basic features of
Altera's FLEX 10K devices which offer quite new design alternatives and
solutions to existing problems than the other CPLDs and FPLDs. A deeper
presentation and analysis of this family is out of scope of this book due to the
fact that the FLEX 10K devices were announced at the time of its completion.
Still, we think that a brief introduction is useful to show a possible trend in the
future CPLDs' development. Altera's FLEX 10K devices are currently
industry's most complex and most advanced CPLDs. Besides logic array blocks
and their logic elements, which are with the same architecture as those in
FLEX8000 devices, FLEX 10K devices incorporate dedicated die areas of
embedded array blocks (EABs) for implementing large specialized functions
providing at the same time programmability and easy design changes. The EAB
consists of memory array and surrounding programmable logic which can easily
be configured to implement required function. Typical functions which can be
implemented in EABs are memory functions or complex logic functions, such as
microcontrollers, digital signal processing functions, data-transformations
functions, and wide data path functions. The LABs are used to implement
general logic.

If the EAB is used to implement memory functions, it provides 2,048 bits,
which are used to create single- or dual-port RAM, ROM or FIFO functions.
When implementing logic, each EAB is equivalent to 100 to 600 gates for
implementation of complex functions, such as multipliers, state machines, or
DSP functions. One FLEX 10K device can contain up to 12 EABs. EABs can be

used independently, or multiple EABs can be combined to implement more complex functions.

The EAB is a flexible block of RAM with registers on the input and output ports. Its flexibility provides implementation of memory of the following sizes: 2,048 x 1, 1,024 x 2, 512 x 4, or 2,048 x 1. This flexibility make it suitable for more than memory, for example by using words of various size as look-up tables and implement functions such as multipliers, error correction circuits, or other complex arithmetic operations. For example, a single EAB can implement a 4 x 4 multiplier with eight inputs and eight outputs providing high performance by fast and predictable access time of the memory block. Dedicated EABs are easy to use, eliminate timing and routing concerns, and provide predictable delays. The EAB can be used to implement both synchronous and asynchronous RAM. In the case of synchronous RAM, the EAB generates its own "write enable" signal and is self-timed with respect to global clock. Larger blocks of RAM are created by combining multiple EABs in "serial" or "parallel" fashion. The global FLEX 10K signals, dedicated clock pins, and EAB local interconnect can drive the EAB clock signals. Because the LEs drive the EAB local interconnect, they can control the "write enable" signal or the EAB clock signal.

In contrast to logic elements which implement very simple logic functions in a single element, and more complex functions in multi-level structures, the EAB implements complex functions, including wide fan-in functions, in a single logic level, resulting in more efficient device utilization and higher performance. The same function implemented in the EAB will often occupy less area on a device, have a shorter delay, and operate faster than functions implemented in logic elements. Depending on its configuration, an EAB can have 8 to 11 inputs and 1 to 8 outputs, all of which can be registered for pipelined designs. Maximum number of outputs depends on the number of inputs. For example, an EAB with 11 inputs can have only 1 output, and an EAB with 8 inputs can have 8 outputs.

Logic functions are implemented by programming the EAB during configuration process with a read-only pattern, creating a large look-up table (LUT). The pattern can be changed and reconfigured during device operation to change the logic function. When a logic function is implemented in an EAB, the input data is driven on the address input of the EAB. The result is looked up in the LUT and driven out on the output port. Using the LUT to find the result of a function is faster than using algorithms implemented in general logic and LEs.

The contents of an EAB can be changed at any time without reconfiguring the entire FLEX 10K device. This enables the change of portion of design while the rest of device and design continues to operate. The external data source used to change the current configuration can be a RAM, ROM, or CPU. For example, while the EAB operates, a CPU can calculate a new pattern for the EAB and reconfigure the EAB at any time. The external data source then downloads the new pattern in the EAB. After this partial reconfiguration process, the EAB is ready to implement logic functions again. If we apply such an design approach that some of the EABs are active and some dormant at the same time, on-the-fly reconfiguration can be performed on the dormant EABs and they can be switched into the working system. This can be accomplished using internal multiplexers to switch-out and switch-in EABs. If the new configuration is stored in external RAM, it has not to be defined in advance. It can be calculated and stored into RAM, and downloaded into the EAB when needed. For example, if the coefficients in an active filter are stored in an EAB, the characteristics of the filter can be changed dynamically by modifying the coefficients. The coefficients are modified by writing to the RAM.

EABs make FLEX 10K devices suitable for a variety of specialized logic applications such as complex multipliers, digital filters, state machines, transcendental functions, waveform generators, wide input/wide output encoders, but also various complex combinatorial functions.

A block diagram of the FLEX 10K architecture is shown in Figure 2.29

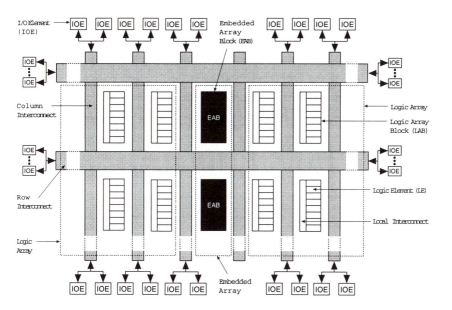

Figure 2.29 FLEX 10K Architecture.

2.3 Xilinx XC4000 FPGAs

The Xilinx XC4000 family of FPGAs provides a regular, flexible, programmable architecture of Configurable Logic Blocks (CLBs) interconnected by a hierarchy of versatile routing resources and surrounded by a perimeter of programmable Input/Output Blocks (IOBs). The devices are customized by loading configuration data into the internal static memory cells (SRAMs).

The basic building blocks used in the Xilinx XC4000 family include

Look-up tables for implementation of logic functions. A designer can use a function generator to implement any Boolean function of a given number of inputs by preloading the memory with the bit pattern corresponding to the truth table of the function. All functions of a function generator have the timing: the time to look up results in the memory. Therefore, the inputs to the function generator are fully interchangeable by simple rearrangement of the bits in the look-up table.

A Programmable Interconnect Point (PIP) is a pass transistor controlled by a memory cell. The PIP is the basic unit of configurable interconnect mechanisms. The wire segments on each side of the transistor are connected depending on the value in the memory cell. The pass transistor introduces resistance into the interconnect paths and hence delay.

A multiplexer is a special case one-directional routing structure controlled by a memory cell. Multiplexers can be of any width, with more configuration bits (memory cells) for wider multiplexers.

The FPGA can either actively read its configuration data out of external serial or byte parallel PROM (master modes) or the configuration data can be written into the FPGA (slave and peripheral modes). FPGAs can be reprogrammed an unlimited number of times allowing the design to change and allowing the hardware to adapt to different user applications.

CLBs provide functional elements for constructing user's logic. IOBs provide the interface between the package pins and internal signal lines. The programmable interconnect resources provide routing paths to connect the inputs and outputs of the CLBs and IOBs to the appropriate networks. Customized configuration is provided by programming internal static memory cells that determine the logic functions and interconnections in the Logic Cell Array (LCA) device. The Xilinx family of FPGAs consists of different circuits with different complexities. Here we present the most advanced type, the Xilinx XC4000. The XC4000 can be used in designs where hardware is changed dynamically, or where hardware must be adapted to different user applications.

2.3.1 Configurable Logic Block

The CLB architecture, shown in Figure 2.30, contains a pair of flip-flops and two independent 4-input function generators. Four independent inputs are provided to each of two function generators which can implement any arbitrarily defined Boolean function of their four inputs. Function generators are labeled F and G.

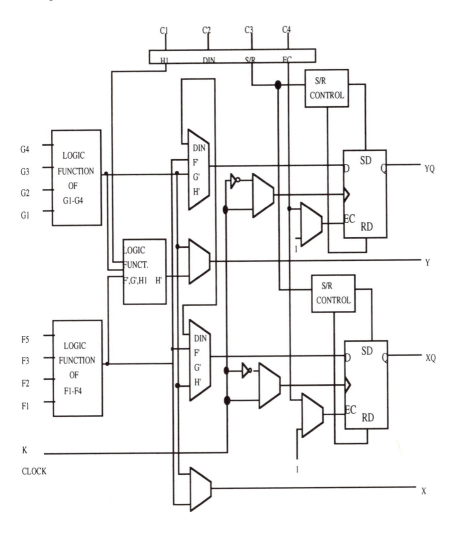

Figure 2.30 CLB Architecture

Function generators are implemented as memory look-up tables (LTUs). A third function generator (labeled H) can implement any Boolean function of its three inputs, two of them being outputs from F and G function generators, and the third input from outside of the CLB. Outputs from function generators are available at the output of the CLB enabling the generation of different combinations of four or five variables Boolean functions. They can even be used to implement some nine variable Boolean functions such as nine-input AND, OR, XOR (parity) or decode in one CLB. The CLB contains two edge-triggered D-type flip-flops with common clock (K) and clock enable (EC) inputs. A Third common input (S/R) can be programmed as either an asynchronous set or reset signal independently for each of the two registers This input can be disabled for either flip-flop. A separate global Set/Reset line (not shown in Figure) is provided to set or reset each register during power up, reconfiguration, or when a dedicated Reset network is driven active. The source of the flip-flop data input can be functions F', G', and H', or the direct input (DIN). The flip-flops drive the XQ and YQ CLB outputs.

Each CLB includes high speed carry logic that can be activated by configuration. As shown in Figure 2.31, two 4-input function generators can be configured as 2-bit adder with built in hidden carry circuitry that is so fast and efficient that conventional speed up methods are meaningless even at the 16-bit level. The fast carry logic opens the door to many new applications involving arithmetic operations, (such as address offset calculations in microprocessors and graphics systems or high speed addition in digital signal processing).

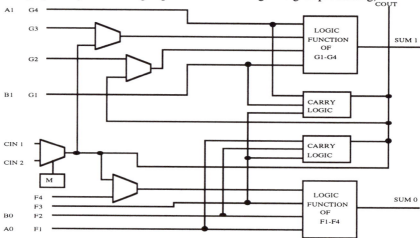

Figure 2.31 Fast Carry Logic in CLB

The Xilinx XC4000 family LCAs include on-chip static memory resources. An optional mode for each CLB makes the memory look-up tables in the function generators usable as either a 16 x 2 or 32 x 1 bit array of Read/Write memory cells, as shown in Figure 2.32. The function generator inputs are used as address bits and additional inputs to the CLB for Write, Enable, and Data-In. Reading memory is the same as using it to implement a function.

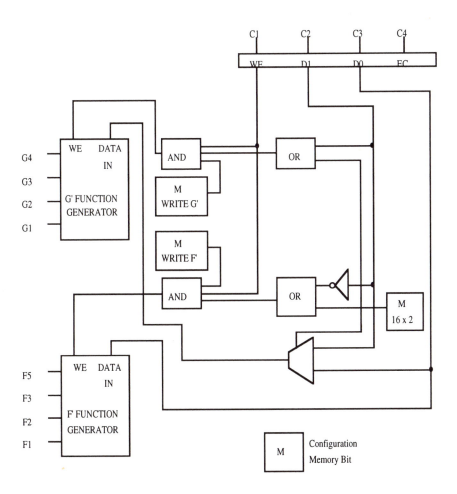

Figure 2.32 Usage of CLB function generators as Read/Write memory cells

The F1-F4 and G1-G4 inputs act as address lines selecting a particular memory cell in each LUT. The functionality of CLB control signals change in this configuration. The H1, DIN, and S/R lines become the two data inputs and Write enable (WE) input for 16 x 2 memory. When the 32 x 1 configuration is selected, D1 acts as the fifth address bit and D0 is the data input. The contents of the memory cell being addressed is available at F' and G' function generator outputs. They can exit through X and Y CLB outputs or can be pipelined using the CLB flip-flops. Configuring the CLB function generators as R/W memory does not affect functionality of the other portions of the CLB, with the exception of the redefinition of control signals.

The RAMs are very fast with read access time being about 5 ns and write time about 6 ns. Both are several times faster than off chip solutions. This opens new possibilities in system design such as registered arrays of multiple accumulators, status registers, DMA counters, LIFO stacks, FIFO buffers, and others.

2.3.2 Input/Output Blocks

User programmable Input/Output Blocks (IOBs) provide the interface between internal logic and external package pins as shown in Figure 2.33. Each IOB controls one package pin. Two lines, labeled I1 and I2 bring input signals into the array. Inputs are routed to an input register that can be programmed as either an edge triggered flip-flop or a level sensitive transparent latch. Each I1 and I2 signals can carry either a direct or registered input signal. By allowing both, the IOB can de-multiplex external signals such as address/data buses, store the address in the flip-flop, and feed the data directly into the wiring. To further facilitate bus interfaces, inputs can drive wide decoders built into the wiring for fast recognition of addresses. Output signals can be inverted or not inverted and can pass directly to the pad or be stored in an edge triggered flip-flop. Optionally, an output enable signal can be used to place the output buffer in a high impedance state, implementing s-state outputs or bidirectional I/O.

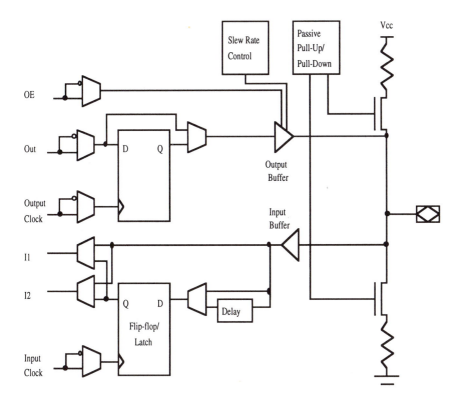

Figure 2.33 IOB architecture

There are a number of other programmable options in the IOB such as programmable pull-up and pull-down resistors, separate input and output clock signals, and global Set/Reset signals as in the case of the CLB.

2.3.3 Programmable Interconnection Mechanism

All internal connections are composed of metal segments with programmable switching points to implement the desired routing. The number of the routing channels is scaled to the size of the array and increases with array size. CLB inputs and outputs are distributed on all four sides of the block as shown in Figure 2.34.

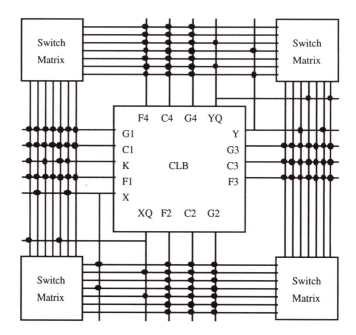

Figure 2.34 Typical CLB connection to adjacent single length lines

There are three types of interconnects distinguished by the relative length of their segments: single length lines, double length lines, and longlines.

The single length lines are a grid of horizontal and vertical lines that intersect at a Switch Matrix between each block. Each Switch Matrix consists of programmable n-channel pass transistors used to establish connections between the single length lines as shown in Figure 2.35. For example, a signal entering on the right side of the Switch matrix can be routed to a single length line on the top, left, or bottom sides, or any combination if multiple branches are required. Single length lines are normally used to conduct signals within localized areas and to provide branching for nets with fanout greater than one.

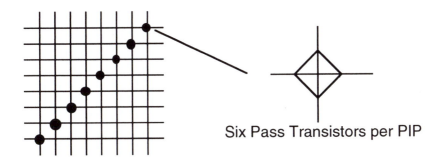

Six Pass Transistors per PIP

Figure 2.35 Switch Matrix

The double length lines, as shown in Figure 2.36 consist of a grid of metal segments twice as long as the single length lines. They are grouped in pairs with the Switch Matrix staggered so each line goes through the Switch Matrix at every other CLB location in that row or column. Longlines form a grid of metal interconnect segments that run the entire length or width of the array (Figure 2.37). Additional long lines can be driven by special global buffers designed to distribute clocks and other high fanout control signals throughout the array with minimal skew. Six of the longlines in each channel are general purpose for high fanout, high speed wiring. CLB inputs can be driven from a subset of the adjacent longlines. CLB outputs are routed to the longlines by way of 3-state buffers or the single interconnect length lines. Communication between longlines and single length lines is controlled by programmable interconnect points at the line intersections, while double length lines can not be connected to the other lines.

A pair of 3-state buffers, associated with each CLB in the array, can be used to drive signals onto the nearest horizontal longlines above and below of the block. The 3-state buffer input can be driven from any X, Y, XQ, or YQ output of the neighboring CLB, or from nearby single length lines with the buffer enable coming from nearby vertical single length lines or longlines. Another 3-state buffer is located near each IOB along the right and left edges of the array. These buffers can be used to implement multiplexed or bi-directional buses on the horizontal longlines. Programmable pull-up resistors attached to both ends of these longlines help to implement a wide wired-AND function.

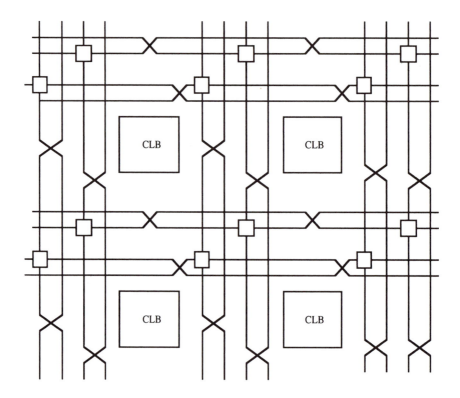

Figure 2.36 Double length lines

The XC4000 family has members with different amounts of wiring for different size ranges. The amount of wire and distribution among different wire lengths is dictated by routability requirements of the FPGAs in the target size range. For the CLB array from 14 x 14 to 20 x 20, each wiring channel includes eight single length lines, four double length lines, six longlines and four global lines. The distribution was derived from an analysis of wiring needs of a large number of existing designs.

All members of the Xilinx family of LCA devices allow reconfiguration to change logic functions while resident in the system. Hardware can be changed as easily as software. Even dynamic reconfiguration is possible, enabling different functions at different times.

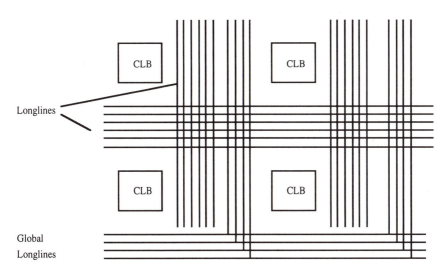

Figure 2.37 Longlines

2.3.4 Device Configuration

Configuration is a process of loading design specific programming data into LCA devices to define the functional operation and the interconnections of internal blocks. This is, to some extent, similar to loading the control registers of a programmable chip. The XC4000 uses about 350 bits of configuration data per CLB and its associated interconnections. Each bit defines the state of a SRAM cell that controls either look-up table bit, multiplexer input, or interconnection pass transistor.

The XC4000 has six configuration modes selected by a 3-bit input code. They are similar to the modes in Altera's family: three are self-loading Master modes, two Peripheral modes, and one is a Serial Slave mode.

The master modes use an internal oscillator to generate the clock for driving potential slave devices and to generate address and timing for external PROM(s) containing configuration data. Data can be loaded either from parallel or serial PROM. In the former case, data is internally serialized into the appropriate format.

Peripheral modes accept byte wide data from a bus. A READY/BUSY status is available as a handshake signal. An externally supplied clock serializes the data. In the Serial Slave mode, the device receives serial configuration data from external source.

LCAs can configure themselves when they sense power up or they can be reconfigured on command while residing in the circuit. A designer can create a system in which the FPGA's program changes during operation.

The LCA can also read back its programming along with the contents of internal flip-flops, latches, and memories. A working part can be stopped and its state recovered. The read-back facility is especially valuable during verification and debugging of prototypes and is also used in manufacturing test.

2.3.5 Designing with XC4000 Devices

As with other FPGAs families, Xilinx FPGAs allow the use of existing design tools for logic or ASIC systems, including schematic entry and hardware description languages. To target an FPGA, a design is passed to FPGA specific implementation software. The interface between design entry and design implementation is a netlist that contains the desired nets, gates, and reference to hard macros.

Although many designs are still done manually, because of the special density and performance requirements, manual designs can be combined with automatic design procedures and can be done completely automatically. Automatic design implementation, the most common method of implementing logic on FPGAs, consists of three major steps: partitioning, placement, and routing.

Partitioning is the separation of the logic into CLBs. It has both a logical and physical component. The connections within a CLB are constrained by the limited intra-block paths and by the limited number of block outputs. The quality of partitioning depends on how well the subsequent placement can be done, so physically related logic should be partitioned into the same block. Placement starts with CLBs, IOBs, hard macros, and other structures in the partitioned netlist. A decision is then made as to which corresponding blocks on the chip should contain those structures. Routing is not as flexible as mask programmed gate arrays. FPGA routing shows very little connectivity between vertical and horizontal segments, requiring many constraints to be taken into account including those for the optimization of the length of nets as well as their delays.

Interactive tools allow constraints on the already known automated algorithms used for MPGAs, postroute improvements on the design, and quick design iterations. The manual editing capability allows users to modify the configuration of any CLB or routing path. In support of an iterative design methodology, Xilinx's automatic place and route system has built-in incremental design facilities. Small changes in a design are incorporated without changing unaffected parts of a design. Large, complex CLBs facilitate incremental changes because a small change can more easily be isolated to a change in a single CLB or a single new routing connection. The incremental change may take only a few minutes, where the original placement and routing may take hours.

3 DESIGN TOOLS AND LOGIC DESIGN WITH FPLDS

This chapter covers aspects of the tools and methodologies used to design with FPLDs. The need for tightly coupled design frameworks, or environments, is discussed and the hierarchical nature of digital systems design is emphasized. All major design description (entry) tools are introduced including schematic entry tools and hardware description languages. The complete design procedure, which includes design entry, processing, and verification, is shown in an example of a simple digital system. An integrated design environment for FPLD-based designs, the Altera's Max+PLUS II environment, is introduced. It includes various design entry, processing, and verification tools.

3.1 Design Framework

FPLD architectures provide identical logic cells (or some of their variations) and interconnection mechanisms as a basis for the implementation of digital systems. These architectures can be used directly for the design of digital circuits. However, the available resources and complexity of designs to be placed in a device require tools that are capable of translating the designer's functions into the exact cells and interconnections needed to form the final design. It is desirable to have design software that will automatically translate designs for different FPLD architectures.

The complexity of FPLDs requires sophisticated design tools that can efficiently handle complex designs. These tools usually integrate several different design steps into a uniform design environment enabling the designer to work with different tools from within the same design framework. They enable design to be performed at a relatively high abstract level, but at the same time allowing the designer to see a physical relationship inside an FPLD device and even change design details at the lowest, physical level.

3.1.1 Design Steps and Design Framework

Design software must perform the following primary functions, as to enable:

Design Entry in some of the commonly used and widely accepted formats. Design entry software should provide an architecture independent design environment that easily adapts to specific designer's needs. The most common design entries belong to the categories of graphic (schematic) design entry, hardware description languages, waveform editors, or some other appropriate tools to transfer designer's needs to a translator.

Translation of design entered by any of the design entry tools or their combinations into the standard internal form that can be further translated for different FPLD architectures. Translation software performs functions such as logic synthesis, timing driven compilation, partitioning, and fitting of design to a target FPLD architecture. Translation mechanisms also provide the information needed for other design tools used in the subsequent design phases.

Verification of a design using functional and timing simulation. In this way many design errors are discovered before actually programming the devices and can be easily corrected using the design entry tools. Usually vendor provided translators produce designs in the forms accepted by industry standard CAE tools that provide extensive verification models and procedures.

Device Programming consisting of downloading design control information into a target FPLD device.

Reusability by providing the libraries of vendor and user designed units that have been proven to operate correctly.

All of the primary functions above are usually integrated into complex design environments or frameworks with a unified user interface. A common element of all these tools is some common circuit representation, most often in the form of so-called netlists.

3.1.2 Compiling and Netlisting

The first step of compiling is the transformation of a design entered in user provided form into the internal form which will be manipulated by the compiler and other tools. A compiler is faced with several issues, the first being will the design fit into the target FPLD architecture at all. Obviously, it depends on the number of input and output pins, but also on the number of internal circuits needed to implement the desired functions. If the design is entered using a graphic editor and the usual schematic notation, a compiler must analyze the possible implementation of all logic elements in existing logic cells of the targeted FPLD. The design is dissected into known three or four input patterns that can be implemented in standard logic cells, and the pieces are subsequently added up. Initially, the compiler has to provide substitutions for the target design gates into equivalent FPLD cells and make the best use of substitution rules. Once substitution patterns are found, a sophisticated compiler eliminates redundant circuitry. This increases the probability that the design will fit into the targeted FPLD device. Compilers translate the design from its abstract form (schematic, equations, waveforms) to a concrete version, a bitmap forming functions and interconnections. An intermediate design form that unifies various design tools is a netlist. After the process of translating a design into the available cells provided by the FPLD (sometimes called the technology mapping phase), the cells are assigned specific locations within the FPLD. This is called cell placement. Once the cells are assigned to a specific locations the signals are assigned to specific interconnection lines. The portion of the compiler that performs placement and routing is usually called a fitter.

A netlist is a text file representing logic functions and their input/output connections. A netlist can describe small functions like flip-flops, gates, inverters, switches, or even transistors. Also it can describe large units (building blocks) like multiplexers, decoders, counters, adders or even microprocessors. They are very flexible because the same format can be used at different levels of description. For example, a netlist with an embedded multiplexer can be rewritten to have the component gates comprising the multiplexer, as an equivalent representation. This is called netlist expansion. One example of the netlist for an 8-to-1 multiplexer is given in the Table 3.1. It simply specifies all gates with their input and output connections, including the inputs and outputs of the entire circuit.

Table 3.1 Example netlist

NETSTART

IN_1	AN4	I(A1, INV_1,INV_2,INV_3) O(IN_1)
IN_2	AN4	I(A2, INV_1,INV_2, S3) O(IN_2)
IN_3	AN4	I(A3, INV_1, S2, INV_3) O(IN_3)
IN_4	AN4	I(A4, INV_1, S2, S3) O(IN_4)
IN_5	AN4	I(A5, S1, INV_2,INV_3) O(IN_5)
INV_1	NOT	I(S1) O(INV_1)
INV_2	NOT	I(S2) O(INV_2)
INV_3	NOT	I(S3) O(INV_3)
B2_1	OR4	I(IN_1, IN_2, IN_3, IN_4) O(B2_1)
OUT_1	OR2	I(B2_1, B2_2) O(Y)

NETEND

| NETIN | A1, A2, A3, A4, A5, S1,S2,S3 |
| NETOUT | Y |

A compiler uses traditional methods to simplify logic designs, but it also uses netlist optimization which represents design minimization after transformation to a netlist. Today's compilers include a large number of substitution rules and strategies in order to provide netlist optimization. One example of a possible netlist optimization of a multiplexer is shown in Figure 3.1. Although simplified, the example shows the basic ideas behind the netlist optimization. Note that five out of eight inputs to a multiplexer are used. The optimizer scans the multiplexer netlist first finding unused inputs, then eliminates gates driven by the unused inputs. In this way a new netlist, without unneeded inputs and gates, is created.

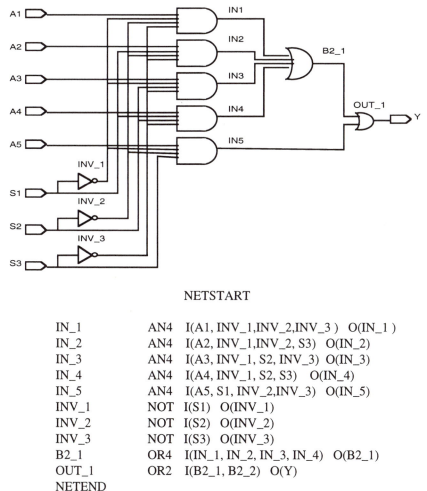

NETSTART

IN_1	AN4	I(A1, INV_1,INV_2,INV_3) O(IN_1)
IN_2	AN4	I(A2, INV_1,INV_2, S3) O(IN_2)
IN_3	AN4	I(A3, INV_1, S2, INV_3) O(IN_3)
IN_4	AN4	I(A4, INV_1, S2, S3) O(IN_4)
IN_5	AN4	I(A5, S1, INV_2,INV_3) O(IN_5)
INV_1	NOT	I(S1) O(INV_1)
INV_2	NOT	I(S2) O(INV_2)
INV_3	NOT	I(S3) O(INV_3)
B2_1	OR4	I(IN_1, IN_2, IN_3, IN_4) O(B2_1)
OUT_1	OR2	I(B2_1, B2_2) O(Y)
NETEND		
NETIN	A1, A2, A3, A4, A5,A6,A7,A8, S1,S2,S3	
NETOUT	Y	

Figure 3.1 Netlist optimization

Even this simple example shows potential payoffs when larger designs are optimized. Optimization procedures are repeated as long as there are gates and flip-flops that can be eliminated or there are logic gates performing identical functions that can be combined and duplication avoided. Even though complex rules are applied during optimization, they save FPLD resources.

After netlist optimization, logic functions are translated into available logic cells with an attempt to map (as much as possible) elementary gates into corresponding logic cells. The next step is to assign logic functions to specific locations within the device. The compiler usually attempts to place them into the simplest possible device if it is not specified in advance.

Cell placement requires iteration and sometimes, if the compiler produces unsatisfactory results, manual placement may be necessary. The critical criteria for cell placement is that interconnections of the cells must be made in order to implement the required logic functions. Additional requirements may be minimum skew time paths or minimum time delays between input and output circuit pins. Usually, several attempts are necessary to meet all constraints and requirements.

Some compilers allow the designer to implement portions of a design manually. Any resource of the FPLD, such as an input/output pin (cell) or logic cell can perform a specific user defined task. In this way; some logic functions can be placed together in specific portions of the device; specific functions can be placed in specific devices (if the projects cannot fit into one device), and inputs or outputs of a logic function can be assigned to specific pins, logic cells, or specific portions of the device. These assignments are taken as fixed by the compiler and it then produces placement for the rest of the design.

Assuming the appropriate placement of cells and other resources, the next step is to connect all resources. This step is called routing. Routing starts with the examination of the netlist that provides all interconnection information, and from inspection of the placement. The routing software assigns signals from resource outputs to destination resource inputs. As the connection proceeds, the interconnect lines become used, and congestion appears. In this case the routing software can fail to continue routing. At this point, the software must replace resource placement into another arrangement and repeat routing again.

As the result of placement and routing design, a file describing the original design is obtained. The design file is then translated to a bitmap that is passed to a device programmer to configure the FPLD. The type of device programmer depends on the type of FPLD programming method (RAM or (E)EPROM based devices).

Good interconnection architectures increase the probability that the placement and routing software will perform the desired task. However, bad routing software can waste a good connection architecture. Even in the case of total interconnectivity, when any cell could be placed at any site and connected to any other site, the software task is very complex. This complexity is increased when constraints are added. Such constraints are, for instance, a timing relationship or requirement that the flip-flops of some register or counter must be placed into adjacent logic cells within the FPLD. These requirements must be met first and then the rest of the circuit is connected. In some cases, placement and routing become impossible. This is the reason to keep the number of such requirements at the minimum.

3.2 Design Entry and High Level Modeling

Design entry can be performed at different levels of abstraction and in different forms. It represents different ways of design modeling, some of them being suitable for behavioral simulation of the system under the design and some being suitable for circuit synthesis. Usually, the two major design entry methods belong to schematic entry systems or textual entry systems.

Schematic entry systems enable a design to be described using primitives in the form of standard SSI and MSI blocks or more complex blocks provided by the FPLD vendor or designer. Textual entry systems use hardware description languages to describe system behavior or structures and their interconnections.

Advanced design entry systems allow combinations of both design methods and the design of subsystems that will be interconnected with other subsystems at a higher level of design hierarchy. Usually, the highest level of design hierarchy is called the project level. Current projects can use and contain designs done in previous projects as its low level design units.

In order to illustrate all design entry methods, we will use an example of a pulse distributor circuit that has Clock as an input and produces five non-overlapping periodic waveforms (clock phases) at the output as shown in Figure 3.2. The circuit has asynchronous Clear, Load initial parallel data, and Enable input which must be active when the circuit generates output waveforms.

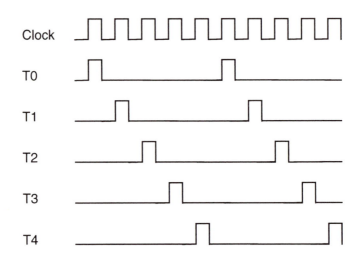

Figure 3.2 Waveforms produced by a pulse distributed circuit

3.2.1 Schematic Entry

Schematic entry is a traditional way to specify a digital system design. A graphics editor is a schematic entry capture program that allows relatively complex designs to be entered quickly and easily. A built-in and extensible primitive and macrofunction libraries provide basic building blocks for constructing a design, while the symbol generation capability enables users to build libraries of custom functions. A graphic editor usually provides a WYSIWYG (What You See Is What You Get) environment. Typical provided primitives include input and output pins, elementary logic gates, buffers, and standard flip-flops. Vendor provided libraries contain macrofunctions equivalent to standard 74- series digital circuits (SSI and MSI), with standard input and output facilities. In the translation process these circuits are stripped off the unused portions such as unused pins, gates, and flip-flops.

A graphic editor enables easy connection of desired output and input pins, editing of new design, duplication of portions or complete design, etc. Symbols can be assigned to new designs and used in subsequent designs. Usual features of Graphic editor are:

- Symbols are connected with single lines or with bus lines. When the name is assigned to a line or bus, it can be connected to another line or bus either graphically or by name only.

- Multiple objects can be selected and edited at the same time.

- Complete areas containing symbols and lines can be moved around the worksheet while preserving signal connectivity. Any selected symbol or area can be rotated.

- Resources can be viewed and edited in Graphic editor such as probes, pins, logic cells, blocks of logic cells, logic and timing assignments.

The example pulse distribution circuit represented by a schematic diagram is shown in Figure 3.3. Some standard 74- series components are used in its implementation.

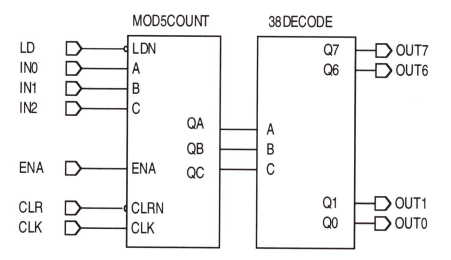

Figure 3.3 Pulse distribution circuit represented by a schematic diagram

3.2.2 Hardware Description Languages

Hardware description languages (HDLs) represent another tool for the description of digital system behavior and structure at different abstraction levels. HDLs belong either to a category of vendor designed languages or general languages that are independent of the vendor.

An example of a vendor provided HDL is Altera's HDL (AHDL). It is a high level, modular language that is integrated into the design environment. AHDL consists of a variety of elements and behavioral statements that describe logic systems. AHDL is a very convenient tool for describing functions such as state machines, truth tables, Boolean functions, conditional logic, and group operations.

It facilitates implementation of combinational logic, such as decoders, multiplexers, arithmetic logic circuits, using Boolean functions and equations, macrofunctions, and truth tables. It allows the creation of sequential logic circuits, such as various types of registers and counters, using Boolean functions and equations, macrofunctions, and truth tables. Frequently used constants and prototypes (of vendor provided or user defined macrofunctions) can be stored in libraries in include files and used where appropriate in new design (textual) files. State machines can be designed using user defined state assignments or by the Compiler.

A detailed introduction to AHDL and a presentation of its features and design mechanisms of digital circuits is given in Chapter 4 and subsequent chapters. For the purpose of providing the "flavor" of the tool, our example pulse distribution circuit is described in AHDL in Example 3.1.

Example 3.1 AHDL Pulse Distribution Circuit.

```
INCLUDE "mod5count";
INCLUDE "38decode";

SUBDESIGN pulsdist
(
        d[2..0]: INPUT
        clk,clr,ld,ena          : INPUT;
        out[4..0] :OUTPUT;
```

```
)

VARIABLE
        counter : mod5count;
        decoder : 8dmux;

BEGIN
        counter.clk = clk;
        decoder.(c,b,a) = counter.(qc,qb,qa);
        out[4..0] = decoder.q[4..0];
END;
```

Include Statements are used to import function prototypes for two already provided user macrofunctions. In the variable section, a variable counter is declared as an instance of the mod5count macrofunction and the variable decoder is declared as an instance of the 38decode macrofunction. They represent a binary modulo-5 counter and decoder of the type 3-to-8, respectively.

The example shows some of the most basic features and potential of hardware description languages. Once described, designs can be compiled and appropriate prototypes and symbols assigned, which can be used in subsequent designs. This approach is not merely leading to the library of designs, but is introducing reusability as a concept in rapid system prototyping.

Although AHDL allows efficient implementation of many combinational and sequential circuits, it can be considered a traditional hardware description language that can be applied mainly to structural and low level digital design.

As a result of the needs and developments in digital systems design methodologies, VHDL (Very High Speed Integrated Circuit Hardware Description Language) has emerged as the standard tool for description of digital systems at various levels of abstraction optimized for transportability among many computer design environments. VHDL is a specification language that follows the philosophy of an object-oriented approach and stresses object-oriented specification and reusability concepts. It describes inputs, outputs, behavior, and functions of digital circuits. It is defined by the IEEE Standard 1076-1987 and revision 1076-1993. In order to compare different design tools on the example, our small pulse distribution circuit is described using VHDL in Example 3.2.

Example 3.2 VHDL Pulse Distribution Circuit.

```
USE work.mycomp.ALL;

ENTITY pulsdist IS
        PORT(d: IN INTEGER RANGE 0 TO 7;
                clk, ld, ena, clr: IN BIT;
                q: OUT INTEGER RANGE 0 TO 255);
END pulsdist;

ARCHITECTURE puls_5 OF pulsdist IS
        SIGNAL a: INTEGER RANGE 0 TO 7;
BEGIN
        cnt_5: mod_5_counter PORT MAP (d,clk,ena,clr,a);
        dec_1: decoder3_to_8 PORT MAP (a, q);
END puls_5;
```

The basic VHDL design units (entity and architecture) appear in this example. The architecture puls_5 of the pulse contains instances of two components from the library mycomp, cnt_5 of type mod_5_counter, and dec_1 of type decoder3_to_8. The complete example of the pulse distributor is presented in the next section, where the hierarchy of design units is introduced. A more detailed introduction to VHDL is presented in Chapter 7. It must be noted that VHDL is a very complex language and as such can be used in a variety of ways. It allows a designer to build his/her own style of design, while still preserving its features of transportability to different design environments.

3.2.3 Hierarchy of Design Units - Design Example

As mentioned earlier, design entry tools usually allow the use of design units specified in a single tool and also the mixing of design units specified in other tools. This leads to the concept of project as the design at the highest level of hierarchy. The project itself consists of all files in a design hierarchy including some ancillary files produced during the design process. The top level design file can be a schematic entry or textual design files, defining how previously designed units, together with their design files are used.

Consider the design hierarchy of our pulse distribution circuit. Suppose the circuit consists of a hierarchy of subcircuits as shown in the Figure 3.4. Figure 3.4 also shows us how portions of the pulse distribution circuit are implemented. At the top of the hierarchy we have the VHDL file (design) that represents the pulse distributor. It consists of modulo-5 counter circuit, denoted MOD-5-COUNTER which is also designed in VHDL. Decoder 3-to-8, denoted DECODER 3-TO-8 is designed using schematic editor. This decoder is designed, in turn, using two 2-to-4 decoders, denoted DECODER 2-TO-4, with enable inputs designed in VHDL. Any of the above circuits could be designed in AHDL, as well. This example just opens the window to a powerful integrated design environment which provides even greater flexibility.

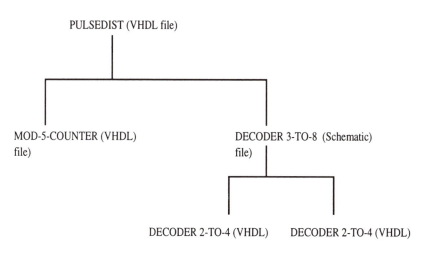

Figure 3.4 Design hierarchy of pulse distribution circuit

The topmost VHDL file specifying our pulse distributor was given in the preceding section. As was mentioned, it represents the structural representation of the circuit that uses two components, `mod_5_counter` and `decoder3_to_8`. VHDL specification of the modulo-5 `counter5` is given in Example 3.3.

Example 3.3 VHDL Module-5 Counter5

```
LIBRARY ieee;
USE ieee.std_logic_1164.ALL;
USE ieee.std_logic_arith.ALL;
USE ieee.std_logic_unsigned.ALL;

ENTITY mod_5_counter IS
        PORT(d: IN INTEGER RANGE 0 TO 4;
                clk, ld, ena, clr: IN BIT;
                q: INOUT INTEGER RANGE 0 TO 4);
END mod_5_counter;

ARCHITECTURE cnt_5 OF mod_5_counter IS
BEGIN
        PROCESS (clk)
                VARIABLE cnt: INTEGER RANGE 0 TO 4;
        BEGIN
                IF(clk'EVENT AND clk='1' THEN
                        IF (clr = '0' OR q = 4) THEN
                                cnt := 0;
                ELSE
                        IF ld = '0' THEN
                                cnt := d;
                                ELSE
                                        IF ena = '1' THEN
                                                cnt := cnt +1;
                                        END IF;
                                END IF;
                        END IF;
                END IF;

                q <= cnt;

        END PROCESS;
    END cnt_4;
```

This counter uses a behavioral style architecture that describes the modulo-5 counter behavior rather than structure. Most VHDL compilers are capable of synthesizing a circuit that carries out the desired function.

Decoder 3-to-8 is designed using a schematic entry with type 2-to-4 decoders, and one standard inverter as its basic components. The schematic diagram of this decoder is shown in Figure 3.5.

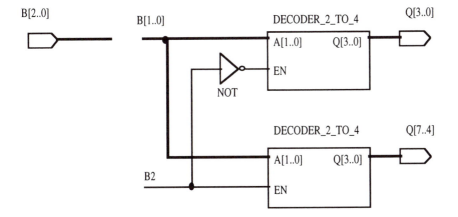

Figure 3.5 Schematic diagram of decoder 3-to-8

Type 2-to-4 decoder is designed using VHDL as shown in Example 3.4.

Example 3.4 VHDL 2-to-4 Decoder.

```
LIBRARY ieee;
USE ieee.std_logic_1164.ALL;

ENTITY decoder_2_to_4 IS
        PORT (a: IN INTEGER RANGE 0 TO 3;
              en: IN BIT;
              q: OUT INTEGER RANGE 0 TO 15;
END decoder_2_to_4;

ARCHITECTURE dec_behav OF decoder_2_to_4 IS

BEGIN
        q <= 1 WHEN (en = '1' AND a = 0) ELSE
             2 WHEN (en = '1' AND a = 1) ELSE
```

```
                    4 WHEN (en = '1' AND a = 2) ELSE
                    8 WHEN (en = '1' AND a = 3) ELSE
                    0;
        END dec_behav;
```

The architecture of the decoder is given again in a behavioral style demonstrating some of the powerful features of VHDL.

Another important component in the hierarchical design of projects is the availability of libraries of primitives, standard 74-series, and application specific macrofunctions, including macrofunctions that are optimized for the architecture of a particular FPLD device or family. Primitives are basic function blocks such as buffers, flip-flops, latches, input/output pins, and elementary logic functions. They are available in both graphical and textual form and can be used in schematic diagrams and textual files. Macrofunctions are high level building blocks that can be used with primitives and other macrofunctions to create new logic designs. Unused inputs, logic gates, and flip-flops are automatically removed by a compiler, ensuring optimum design implementation. Macrofunctions are usually given together with their detailed implementation, enabling the designer to copy them into their own library and edit them according to specific requirements.

The common denominator of all design entry tools is the netlist level at which all designs finally appear. If standard netlist descriptions are used, then further tools that produce actual programming data or perform simulation can be specified by another design specification tool, provided the compiler produces standard netlist formats.

3.3 Design Verification and Simulation

Design verification is necessary because there are bugs and errors in the translation, placement and routing processes, as well as errors made by a designer. Most verification tools are incorporated into design tools and examine netlists and analyze properties of the final design. For instance, a design checker can easily identify the number of logic cells driven by any logic cell and determine how it contributes to a cumulative load resulting in a time delay attached to the driving cell's output. If the delay is unacceptable, the designer must split the load among several identical logic cells. Similarly, a design checker can identify unconnected inputs of logic cells which float and produce noise problems.

While some checks can be performed during design compilation, many checks can only be done during a simulation that enables assessing functionality and the timing relationship and performance of an FPLD based design.

Regardless of the type of logic simulation, a model of the system is created and driven by a model of inputs called stimuli or input vectors, that generate a model of output signals called responses or output vectors. Simulation is useful not only for observing global behavior of the system under design, but also because it permits observation of internal logic at various levels of abstraction, which is not possible in actual systems.

Two types of simulation used in digital systems design are functional simulation and timing simulation. Functional simulation enables observation of design units at the functional level, by combining models of logic cells with models of inputs to generate response models that takes into account only relative relationships among signals and neglecting circuit delays. This type of simulation is useful for a quick analysis of system behavior, but produces inaccurate results because propagation delays are not taken into account.

Timing simulation takes into account additional element in association with each cell model output, a time delay variable. Time delay enables more realistic modeling of logic cells and the system as the whole. They consist of several components that can or cannot be taken into account, such as time delay of logic cells, without considering external connections, time delays associated with the routing capacitance of the metal connecting outputs with the inputs of logic cells, and time delay which is the function of the driven cell input impedances.

Timing simulation is the major part of verification process of the FPLD design. Timing simulator uses two basic information to produce output response, Input Vectors and Netlists. Input vectors are given in either tabular or graphical form. Input timing diagrams represent a convenient form to specify stimuli of simulated design.

Netlists represent an intermediate form of the system modeled. Besides connectivity information, netlists contain information about delay models of individual circuits and logic cells, as well as logic models that describe imperfections in the behavior of logic systems. These models increase complexity of used logic, but at the same time improve quality of the model and system under design.

The simulator applies the input vectors to the system model under design and after processing according to the input netlists and models of individual circuits, produces the resulting output vectors. Usually, outputs are presented in the form of timing diagrams. Later, both input and output timing diagrams can be used by electronic testers to compare simulated and real behavior of the design.

In order to perform simulation, the simulator has to maintain several internal data structures that easily and quickly help find the next event requiring the simulation cycle to start. A simulation event is the occurrence of a netlist node (gate, cell, output, etc.) making a binary change from one value to another. The scheduler is a part of the simulator that keeps a list of times and events, and dispatches events when needed. The process initiates every simulated time unit regardless of an event existence (in that case we say that simulation is time driven) or only at the time units in which there are some events (event driven simulation).

The second type of simulation is more popular and more efficient in today's simulators. The most important data structure is the list of events which must be ordered according to increased time of occurrence.

In the ideal case, these changes are simply binary (from zero to one and vice versa), but more realistic models take into account imperfect or sometimes unspecified values of signals. The part of a simulator called the evaluation module is activated at each event and uses models of functions that describe behavior of the subsystems of design. These models take into account more realistic electrical conditions of circuit behavior, such as three-state outputs, unknown states, and time persistence, essentially introducing multi-valued instead of common binary logic. Some simulators use models with up to twelve values of the signals. This leads to complex truth tables and complex and time consuming simulation even for simple logic gates, but also produces a more accurate simulation results.

3.4 Integrated Design Environment Example: Altera's Max+PLUS II

An integrated design environment for EPLD/CPLD design represents a complete framework for all phases of the design process, starting with design entry and ending with device programming. Altera's Max+PLUS II is an integrated software package for designing with Altera programmable devices. The same design can be retargeted to various devices without changes to the design itself. Max+PLUS II consists of a spectrum of logic design tools and capabilities, such as a variety of design entry tools for hierarchical projects, logic synthesis algorithms, timing-driven compilation, partitioning, functional and timing simulation, linked multi device simulation, timing analysis, automatic error location, and device programming and verification. It is also capable of reading netlist files produced by other vendor systems or producing netlist files for other industry standard CAE software.

The Max+PLUS II design environment is shown in Figure 3.6. The heart of the environment is a compiler capable of accepting design specifications in various design entry tools, and producing files for two major purposes, design verification and device programming. Design verification is performed using functional or timing simulation, or timing analysis and device programming, is performed by Altera's or other industry standard programmers. Output files produced by the Max+PLUS II compiler can be used by other CAE software tools.

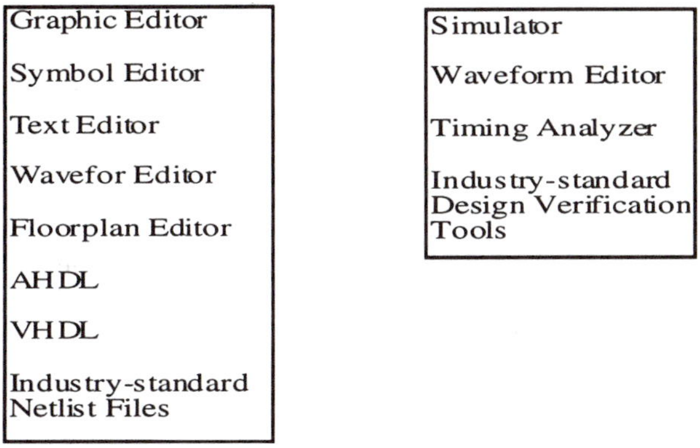

Graphic Editor
Symbol Editor
Text Editor
Wavefor Editor
Floorplan Editor
AHDL
VHDL
Industry-standard Netlist Files

Simulator
Waveform Editor
Timing Analyzer
Industry-standard Design Verification Tools

Figure 3.6 Max+PLUS II Design Environment.

Once the logic design is created, the entity is called a project. A project can include one or more subdesigns (previously designed projects). It combines different types of subdesigns (files) into a hierarchical project, choosing the design entry format that best suits each functional block. In addition, large libraries of Altera provided macrofunctions simplify design entry. Macrofunctions are available in different forms and can be used in all design entry tools.

A project consists of all files in a design hierarchy. If needed, this hierarchy can be displayed at any moment, and the designer sees all files that make the project, including design and ancillary files. Design files represent a graphic, text, or waveform file created with a corresponding editor, or with another industry standard schematic text editor or a netlist writer. The Max+PLUS II compiler can process the following files:

- Graphic design files (.gdf)

- Text design files (.tdf)

- Waveform files (.wdf)

- VHDL files (.vhd)

- OrCAD schematic files (.sch)

- EDIF input files (.edf)

- Xilinx netlist format files (.xnf)

- Altera design files (.adf)

- State machine files (.smf)

Ancillary files are associated with a project, but are not part of a project hierarchy tree. Most of them are generated by different Max+PLUS II functions and some of them can be entered or edited by a designer. Examples of ancillary files are assignment and configuration files (.acf) and report files (.rpt).

The Max+PLUS II compiler provides powerful project processing and customization to achieve the best or desired silicon implementation of a project. Besides fully automated procedures, it allows designer to perform some of the assignments or functions manually to control a design. A designer can enter, edit, and delete resource and device assignments that control project compilation, including logic synthesis, partitioning, and fitting.

The Max+PLUS II design environment is Windows based. This means that all functions can be invoked using menus or simply by clicking on different buttons with the icons describing corresponding functions.

3.4.1 Design Entry

Max+PLUS II provides three design entry editors: the graphic, text, and waveform editors. Two additional editors are included to help facilitate design entry, the floorplan and symbol editor.

Design entry methods supported by Max+PLUS II are:

- Schematic, designs are entered in schematic form using Graphic editor.

- Textual, AHDL or VHDL designs are entered using Altera's or any other standard text editor

- Waveform, designs are specified wit Altera's waveform editor

- Netlist, designs in the form of netlist files or designs generated by other industry standard CAE tools can be imported into Max+PLUS II design environment

- Pin, logic cell, and chip assignments for any type of design file in the current project can be entered in a graphical environment with the floorplan editor.

- Graphic symbols that represent any type of design file can be generated automatically in any design editor. Symbol editor can be used to edit symbols or create own customized symbols.

The Assign menu, accessed in any Max+PLUS II application, allows the user to enter, edit, and delete the types of resource and device assignments that control project compilation. This information is saved in assignment and configuration files (.acf) for the project. Assignment of device resources can be controlled by the following types of assignments:

- Clique assignments specify which logic functions must remain together in the same logic array block, row, or device

- Chip assignments specify which logic must remain together in a particular device when a project is partitioned into multiple devices

- Pin assignments assign the input or output of a single logic function to a specific pin, row, or column within a chip

- Logic cell assignments assign a single logic function to a specific location within a chip (to a logic cell, I/O cell, LAB, row, or column)

- Probe assignments assign a specific name to an input or output of a logic function

- Connected pin assignments specify how two or more pins are connected externally on the printed circuit board.

- Device assignments assign project logic to a device (for example, maps chip assignments to specific devices in multi-device project)

- Logic option assignments that specify the logic synthesis style in logic synthesis (synthesis style can be one of three Altera provided or specified by designer)

- Timing assignments guides logic synthesis tools to the desired performance for input to non-registered output delays (t_{PD}), clock to output delays (t_{CO}), clock setup time (t_{SU}), and clock frequency (f_{MAX}).

Max+PLUS II allows preservation of the resource assignments the compiler made during the most recent compilation so that we can produce the same fit with subsequent compilation. This feature is called back annotation. It becomes essential because after compiling all time delays are known and the design software can calculate a precise annotated netlist for the circuit by altering the original netlist. The subsequent simulation using this altered netlist is very accurate and can show trouble spots in the design that are not otherwise observable.

Some global device options can be specified before compilation such as the reservation of device capacity for future use or some global settings such as an automatic selection of a global control signal like the Clock, Clear, Preset, and Output Enable. The compiler can be directed to automatically implement logic in I/O cell registers.

3.4.2 Design Processing

Once a design is entered, it is processed by the Max+PLUS II compiler producing the various files used for verification or programming. The Max+PLUS II compiler consists of a series of modules that check a design for errors, synthesize the logic, fit the design into the needed number of Altera devices, and generate files for simulation, timing analysis, and device programming. It also provides a visual presentation of the compilation process, showing which of the modules is currently active and allowing this process to be stopped.

Besides design entry files, the inputs to the compiler are the assignment and configuration files of the project (.acf), symbol files (.sym) created with Symbol editor, include files (.inc) imported into text design files containing function prototypes and constants declarations, and library mapping files (.lmf) used to map EDIF and OrCAD files to corresponding Altera provided primitives and macrofunctions.

The compiler netlist extractor first extracts information that defines hierarchical connections between a project's design files and checks the project for basic design entry errors. It converts each design file in the project into a binary Compiler Netlist File (.cnf) and creates one or more Hierarchy Interconnect Files (.hif), a Symbol File (.sym) for each design file in a project, and a single Node Database File (.ndb) that contains project node names for assignment node database.

If there are no errors, all design files are combined into a flattened database for further processing. Each Compiler Netlist File is inserted into the database as many times as it is used in the original hierarchical project. The database preserves the electrical connectivity of the project.

The compiler applies a variety of techniques to implement the project efficiently in one or more devices. The logic synthesizer minimizes logic functions, removes redundant logic, and implements user specified timing requirements.

If a project does not fit into a single device, the partitioner divides the database into the minimal number of devices from the same device family. A project is partitioned along logic cell boundaries and the number of pins used for inter-device communication is minimized.

The fitter matches project requirements with known resources of one or more device. It assigns each logic function to a specific logic cell location and tries to match specific resource assignments with available resources. If it does not fit, the fitter issues a message with the options of ignoring some or all of the required assignments.

Regardless if a fit is achieved or not, a report file (.rpt) is created showing how a project will be implemented. It contains information on project partitioning, input and output names, project timing, and unused resources for each device in the project.

At the same time, the compiler creates a functional or timing simulation netlist file (.snf) and one or more programming files that are used to program the devices. The programming image can be in the form of one or more programmer object files (.pof), or SRAM object files (.sof files). For some devices, JEDEC files (.jed) can be generated.

As an example, our pulse distributor circuit is compiled by the Max+PLUS II compiler without constraints or user required assignments of resources. The tables below show some of the results of compilation. The compiler has placed the pulsdist circuit into the EPF8282LC84 device with the logic cell utilization of 5%. Other important information about the utilization of resources is available in the tables 3.2 through 3.5 given below.

Table 3.2 Device summary for pulsdist circuit

Chip	Device	Input pins	Output pins	Bidir pins	Logic cells	% utilized
Pulsdist	EPF8282 LC84	7	8	0	12	5%
User pins		7	8			

Table 3.3 Logic cell utilization

Column Row	01	02	03	04	05	06	07	08	09	10	11	12	13	Total
A	0	0	0	0	0	0	0	0	0	0	0	0	0	0
B	4	3	1	1	1	1	1	0	0	0	0	0	0	12
Total	4	3	1	1	1	1	1	0	0	0	0	0	0	12

Table 3.4 Interconnect mechanism usage

LAB	Logic cells	Column Interconn. Driven	Row Interconn. Driven	Clocks	Clear/ Preset	External Interconn.
B1	4/8 (50%)	0/8 (0%)	3/8 (37%)	1/2	1/2	6/24 (25%)
B2	3/8 (37%)	0/8 (0%)	3/8 (37%)	0/2	0/2	3/24 (12%)
B3	1/8 (12%)	0/8 (0%)	1/8 (12%)	0/2	0/2	3/24 (12%)
B4	1/8 (12%)	0/8 (0%)	1/8 (12%)	0/2	0/2	3/24 (12%)
B5	1/8 (12%)	0/8 (0%)	1/8 (12%)	0/2	0/2	3/24 (12%)
B6	1/8 (12%)	0/8 (0%)	1/8 (12%)	0/2	0/2	3/24 (12%)
B7	1/8 (12%)	0/8 (0%)	1/8 (12%)	0/2	0/2	3/24 (12%)

Table 3.5 Other resources utilization

Total dedicated input pins used	1/4 (25%)
Total I/O pins used	16/64 (25%)
Total logic cells used	12/208 (5%)
Total input pins required	7
Total input registers required	0
Total output pins required	8
Total output registers required	0
Total buried I/O cell registers required	0

Total bidirectional pins required	0
Total reserved pins required	2
Total logic cells required	12
Total flip-flops required	3
Total logic cells in carry chains	3
Total number of carry chains	1
Total logic in cascade chains	0
Total number of cascade chains	0
Synthesized logic cells	0/208 (0%)

3.4.3 Design Verification

The process of project verification is aided with two major tools: the simulator, and the timing analyzer. The simulator tests the logical operation and internal timing of a project. To simulate a project, a Simulator Netlist File (.snf) must be produced by the compiler. An appropriate SNF file (for functional, timing, or linked multi-project simulation) is automatically loaded when the simulator is invoked.

The input vectors are in the form of a graphical waveform Simulator Channel File (.scf) or an ASCII Vector File (.vec). The Waveform editor creates a default SCF file. The simulator allows the designer to check the outputs of the simulation against any outputs in SCF, such as user defined outputs or outputs from a previous simulation. It can also be used to monitor glitches, oscillations, and setup and hold time violations.

An example of the simulator operation is given for our pulse distributor circuit in Figure 3.7. Input vectors are denote by capital letter I, and output vectors by capital letter O. A total of 800 ns was simulated. In Figure 3.7 a 280 ns interval of the simulation is shown.

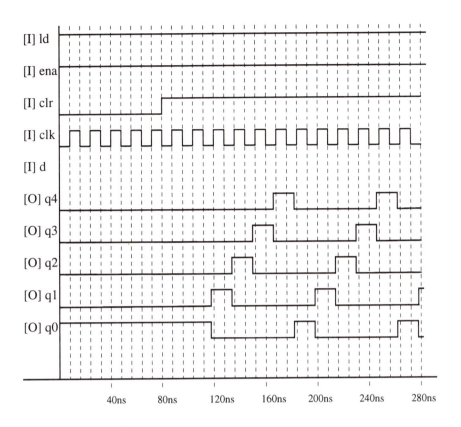

Figure 3.7 Simulation results for pulse distributor circuit

The input clock period is 16 ns. Further timing analysis has shown that the circuit can safely run to the minimum clock period of 15.7 ns or frequency of 63.69 MHz.

The Max+PLUS II timing analyzer allows the designer to analyze timing performance of a project after it has been optimized by the compiler. All signal paths in the project can be traced, determining critical speed paths and paths that limit the project's performance. The timing analyzer uses the network and timing information from a timing Simulator Netlist File (.snf) generated by the compiler. It generates three types of analyses, the delay matrix, the set up/ hold matrix, and the registered performance display.

The delay matrix shows the shortest and longest propagation delay paths between multiple source and destination nodes.

The setup/hold matrix shows the minimum required setup and hold times from input pins to the D, Clock, and latch Enable inputs to flip-flops and latches.

The registered performance display shows the results of a registered performance analysis, including the performance limited delay, minimum Clock period, and maximum circuit frequency.

After the timing analyzer completes an analysis, it is possible to select a source or destination node and list its associated delay paths. Using the message processor it is easy to open and list the paths for the selected node and locate a specific path in the original design file.

3.4.4 Device Programming

The last portion of Altera's integrated design environment is the hardware and software necessary for programming and verifying Altera devices. The software part is called the Max+PLUS II programmer. For EPROM base devices Altera provides an add-on Logic Programmer card (for PC-AT compatible computers) that drives the Altera Master Programming Unit (MPU). The MPU performs continuity checks to ensure adequate electrical contact between the programming adapter and the device. With the appropriate programming adapter, the MPU also supports functional testing. It allows the application of simulation input vectors to verify its functionality.

For the FLEX 8000 family, Altera provides the FLEX download cable and the BitBlaster. The FLEX download cable can connect any configuration EPROM programming adapter, which is installed on the MPU, to a single target FLEX 8000 device. The BitBlaster serial download cable is a hardware interface to a standard RS-232 port that provides configuration data to FLEX 8000 devices. The BitBlaster allows the designer to configure the FLEX 8000 device independently from the MPU or any other programming hardware.

4 INTRODUCTION TO DESIGN USING AHDL

This chapter is devoted to design using Altera's Hardware Description Language (AHDL). The basic features of AHDL are introduced without a formal presentation of the language. Small examples are given to illustrate its features and usage. The design of combinatorial logic in AHDL including the implementation of bidirectional pins, standard sequential circuits such as registers and counters, and state machines is presented. The implementation of user designs as hierarchical projects consisting of a number of subdesigns is also shown.

4.1. AHDL Design Entry

The Altera Hardware Description Language (AHDL) is a high level, modular language especially suited for complex combinatorial logic, group operations, state machines, and truth tables, and recently extended with features that facilitate design structures and more complex projects. AHDL Text Design Files (.tdf) can be entered using any text editor, and subsequently compiled and simulated, and are used to program Altera FPLDs.

AHDL allows a designer to create hierarchical designs (projects) which also incorporate other types of design files. AHDL design entry and relationship to other types of design files within MAX+PLUS II design environment are illustrated in Figure 4.1. A symbolic representation of a TDF is automatically created and can be incorporated into a Graphic Design File (.gdf). Also, custom functions, as well as provided macrofunctions can be incorporated into any TDF. Altera provides Include Files (.inc) with function prototypes for all provided functions in the macrofunction library. A hierarchical project can contain TDFs, GDFs, and EDIF Input Files (.edf) at any level of the project hierarchy. Waveform Design Files (.wdf), Altera Design Files (.adf), and State Machine Files (.smf) can be used only at the lower level of a project hierarchy.

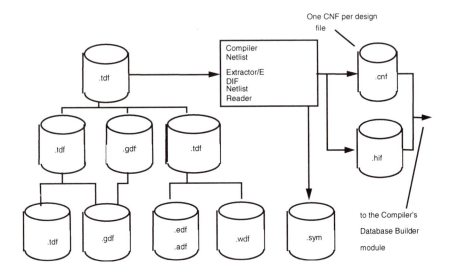

Figure 4.1 AHDL Design Entry and Relationship to MAX+PLUS II Design Environment

4.1.1 Text Design Files

A TDF must contain either a Design Section or a Subdesign Section/Logic Section combination, or both. The sections and statements that appear in TDF (in the order of appearance) are:

- **Title Statements** (Optional) provide comments for the Report File (.rpt) generated by the MAX+PLUS II Compiler.

- **Include Statements** (optional) specify an Include File that replaces the Include Statement in the TDF.

- **Constant Statements** (optional) specify a symbolic name that can be substituted for a constant.

- **Define Statement** (optional) defines an evaluated function, which is a mathematical function that returns a value that is based on optional argument.

- **Parameters Statement** (optional) declares one or more parameters that control the implementation of a parameterized functions. A default value can be specified for each parameter.

- **Function Prototype Statements** (optional) declare the ports of a macrofunction or primitive and the order in which these ports must be declared in any in-line reference. In parameterized functions, it also declares the parameters used by the function.

- **Options Statements** (optional) set the Turbo and Security Bits of Altera devices and specifies logic options and logic synthesis styles. This statement can be placed before the Design Section, inside the Design Section, or inside the Device Specification. In the newer versions of the MAX+PLUS II environment, various options are not specified in TDF, but rather they are set using specialized menus and windows for that purpose.

- **Assert Statement** (optional) allows the designer to test validity of an arbitrary expression and report the results.

- **Subdesign Sections** (required) declare the input, output, and bidirectional ports of an AHDL TDF.

- **Variable Sections** (optional) declare variables that represent and hold internal information. Variables can be declared for ordinary or three-state nodes, primitives, functions and state machines.

- **Logic Sections** (required) define the logical operations of the file. This section can define logic with Boolean expressions, conditional logic, and truth tables. It also supports conditional and iterative logic generation, and the capability to test the validity of an arbitrary expressions and report the results.

AHDL is a concurrent language. All behavior specified in the Logic Section of a TDF is evaluated at the same time. Equations that assign multiple values to the same AHDL node or variable are logically connected (ORed if the node or variable is active high, ANDed if it is active low). The Design Section contains an architectural description of the TDF. The last entries in the TDF, the Subdesign Section, Variable Section (optional), and Logic Section, collectively contain the behavioral description of the TDF.

If used, macrofunctions are connected through their input and output ports to the design file at the next higher level of the hierarchy. The contents of the Include File, an ASCII file, are substituted wherever an include Statement is found in the TDF. It is recommended to include only Constants or Function Prototype Statements in the Include File.

When the TDF is entered using a text editor, its syntax can be checked with the Save & Check command, or all files can be compiled in a project with the Save & Compile command. The MAX+PLUS II compiler automatically generates a symbol for the current file, which can be used in GDF. After the project has compiled successfully, you can perform optional simulation and timing analysis, and then program one or more devices.

4.2. AHDL Basics

The Altera Hardware Description Language is a text entry language for describing logic designs. It is incorporated into the MAX+PLUS II design environment. AHDL consists of a variety of elements that are used in behavioral statements to describe logic.

4.2.1 Using Numbers and Constants

Numbers are used to specify constant values in Boolean expressions and equations. AHDL supports all combinations of decimal, binary, octal, and hexadecimal numbers.

Example 4.1 is of an address decoder that generates an active-high chip enable when the address is FF30 (Hex) or FF50 (Hex) present on the input.

Example 4.1 Address Decoder

```
SUBDESIGN decode
 (
```

```
            address[15..0]                   :INPUT;
            chip_enable1,chip_enable2        :OUTPUT;
)
BEGIN

            chip_enable1 = (address[15..0] == H"FF30");
            chip_enable2 = (address[15..0] == H"FF50");
END;
```

The decimal numbers 15 and 0 are used to specify bits of the address bus. The hexadecimal numbers H"FF30" and H"FF50" specify the addresses that are decoded. Example program can be stored in the TDF. The equivalent GDF file is shown in Figure 4.2.

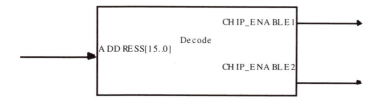

Figure 4.2 GDF equivalent of the decoder circuit

Constants can be used to give a descriptive name to a number. This name can be used throughout a file. In the case that the change of the value of a constant is needed, it is done at only one place, where constant is declared.

In Example 4.1 above, we can introduce the constants IO_ADDRESS1 and IO_ADRESS2 to describe the addresses that are to be decoded. The new TDF is shown in Example 4.2.

Example 4.2 TDF of Modified Decoder.

```
CONSTANT IO_ADDRESS1 = H"FF30";
CONSTANT IO_ADDRESS2 = H"FF50";

SUBDESIGN decode1
(
        a[15..0]                 :INPUT;
        ce1, ce2                 :OUTPUT;
)
```

```
BEGIN
        ce1 = (a[15..0] == IO_ADDRESS1);
        ce2 = (a[15..0] == IO_ADDRESS2);

END;
```

Constants can be declared using arithmetic expressions which include the other already declared constants. The Compiler evaluates arithmetic expressions and replaces them by numerical values.

For example, we can declare constant:

```
CONSTANT  IO_ADDRESS1 = H"FF30";
```

and then

```
CONSTANT  IO_ADDRESS2 = IO_ADDRESS1 + H"0010";
CONSTANT  IO_ADDRESS3 = IO_ADDRESS1 + H"0020";
```

using address H'FF30" as a base address, and generating other addresses relative to the base address.

Another example of using constants is to substitute the values for parameters during compilation. The design can contain one or more parameters whose values are replaced with actual values at the compilation time. For example, the target device family can be a parameter, and the actual value of parameter can be substituted by a constant:

```
PARAMETERS
(
DEVICE_FAMILY
);

CONSTANT FAMILY1 = "MAX7000";
CONSTANT FAMILY2 = "FLEX8000";
```

which is further used within the Subdesign Section to compile the design for a specific device family depending on the value of DEVICE_FAMILY parameter (which can be FAMILY1 or FAMILY2).

4.2.2 Combinatorial Logic

Combinatorial logic is implemented in AHDL with Boolean expressions and equations, truth tables, and a variety of macrofunctions. Boolean expressions are sets of nodes, numbers, constants, and other Boolean expressions separated by operators and/or comparators, and optionally grouped with parentheses. A Boolean equation sets a node or group equal to the value of a Boolean expression. Example 4.3 shows simple Boolean expressions that represent logic gates.

Example 4.3 Boolean Expressions for Logic Gates.

```
SUBDESIGN bool1
(
        a0, a1, b0, b1              :INPUT;
        s0, s1                     :OUTPUT;
)
BEGIN
        s0 = a0 & a1 & !b1;
        s1 = s0 # b0;
END;
```

Since equations are evaluated concurrently, their order in the file above is not important. The GDF equivalent of the above TDF is shown in Figure 4.3.

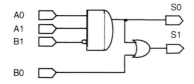

Figure 4.3 GDF Representation of the Circuit

4.2.3 Declaring Nodes

A node is declared with a node declaration in the Variable Section. It can be used to hold the value of an intermediate expression. Node declarations are useful when a Boolean expression is used repeatedly. The Boolean expression can be replaced with a more descriptive node name. Example 4.4 below performs the same function as the former one, but it uses a node declaration which saves device resources if repeatedly used.

Example 4.4 Boolean Expressions with Node Declarations.

```
SUBDESIGN bool2
(
        a0, b0, a1, b1        :INPUT;
        s1                    :OUTPUT;
)
VARIABLE
        inter                 :NODE;
BEGIN
        inter = a0 & a1 & !b1;
        s1 = inter # b0;
END;
```

GDF equivalent to this TDF is shown in Figure 4.4.

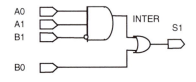

Figure 4.4 GDF Representation of the Node Declared Circuit

4.2.4 Defining Groups

A group, which can include up to 256 members (bits), is treated as a collection of nodes and is acted upon as one unit. In Boolean equations, a group can be set equal to a Boolean expression, another group, a single node, Vcc, GND, 1, or 0. In each case, the value of the group is different. Once the group has been defined, [ ] is a shorthand way of specifying the entire range. A subgroup is represented using a subrange of the indices. The Example 4.5 is of legal group values.

Example 4.5 Legal Group Values.

```
a[7..0] = b[15..8]; % a7 connected to b15, a6 to b14,...%
a[2..0] = inter;    % all bits connected to inter %
a[7..0] = Vcc;      % all bits connected to Vcc %
```

The Options Statement can be used to specify the most significant bit (MSB) or the least significant bit (LSB) of each group. For example

```
OPTIONS BIT0 = MSB;
```
or

```
OPTIONS BIT0 = LSB;
```

specify the lowest numbered bit (bit 0) to be either MSB or LSB, respectively.

4.2.5 Conditional Logic

Conditional logic chooses from different behaviors depending on the values of the logic inputs. AHDL provides two statements for conditional logic implementation, the IF statement and the Case statement.

• **IF statements** evaluate one or more Boolean expressions, then describe the behavior for different values of the expression. IF statement can be in the simple IF THEN form or in any variant of IF THEN ELSIF... ELSE forms.

• **Case statements** list alternatives that are available for each value of an expression. They evaluate expression, then selects a course of action on the basis of the expression value.

Example 4.6 is of the use of IF Statement in a priority encoder and is given below.

Example 4.6 IF Statement Use.

```
SUBDESIGN priority
(
        prior4,prior3,prior2,prior1        :INPUT;
        prior_code[2..0]                   :OUTPUT;
)
BEGIN
        IF prior4 THEN
                prior_code[] = 4;
        ELSIF prior3 THEN
                prior_code[] = 3;
        ELSIF prior2 THEN
                prior_code[] = 2;
        ELSIF prior1 THEN
                prior_code[] = 1;
        ELSE
                prior_code[] = 0;
        ENDIF;
END;
```

The inputs `prior4`, `prior3`, `prior2`, and `prior1` are evaluated to determine whether they are driven by Vcc. The IF Statement activates the equations that follow the highest priority IF or ELSE clause that is active. Output priority code is represented by 3-bit value. If no input is active, then the output code 0 is generated.

Example 4.7 shows the use of a Case Statement in specifying the 2-to-4 decoder that converts two bit code into "one hot" code.

Example 4.7 Case Statement Use.

```
SUBDESIGN decoder_2_to_4
(
        inpcode[1..0] :INPUT;
        outcode[3..0] :OUTPUT;
)
BEGIN
        CASE inpcode[ ] IS
                WHEN 0 => outcode[ ] = B"0001";
                WHEN 1 => outcode[ ] = B"0010";
```

```
                        WHEN 2 => outcode[ ] = B"0100";
                        WHEN 3 => outcode[ ] = B"1000";
            END CASE;
    END;
```

The input group inpcode[1..0] has the value 0, 1, 2, or 3. The equation following the appropriate => symbol is activated.

It is important to know that besides similarities, there are also differences between IF and Case Statements. Any kind of Boolean expression can be used in an IF Statement, while in a Case Statement, only one Boolean expression is compared to a constant in each WHEN clause.

4.2.6 Decoders

A decoder contains combinatorial logic that converts input patterns to output values or specifies output values for input patterns. Truth Table Statements can be used to create a decoder. Example 4.8 illustrates this.

Example 4.8 Truth Table Decoder.

```
    SUBDESIGN decoder
    (
            inp[1..0]      :INPUT;
            a, b, c, d     :OUTPUT;
    )
    BEGIN
            TABLE
                    inp[1..0] => a, b, c, d;

                    H"0"          => 1, 0, 0, 0;
                    H"1"          => 0, 1, 0, 0;
                    H"2"          => 0, 0, 1, 0;
                    H"3"          => 0, 0, 0, 1;
            END TABLE;
    END;
```

In Example 4.8, the output pattern for all 4 possible input patterns of inp[1..0] is described in the Truth Table Statement.

In the case that decoder is not partial one (not decoding all possible input combinations), the Default Statement can be used to specify the output of the decoder when not-specified values of input appear as shown in Example 4.9:

Example 4.9 Use of the Default Statement to Specify the Decoder Output

```
SUBDESIGN partial
(
inpcode[3..0]:  INPUT;
outcode[4..0]: OUTPUT;
)
BEGIN
        DEFAULTS
                outcode[]= B"11111"; %value of output for%
                                     %unspecified input codes%
        END DEFAULTS;

        TABLE
                inpcode[]       => outcode[];

                B"0001"         => B"01000";
                B"0011"         => B"00100";
                B"0111"         => B"00010";
                B"1111"         => B"00001";
        END TABLE;
END;
```

Example 4.10 represents an address decoder for a generalized 16 bit microprocessor system.

Example 4.10 Address Decoder for a 16-Bit Microprocessor.

```
SUBDESIGN decode2
(
        addr[15..0], m/io                    :INPUT;
        rom, ram, print, sp[2..1]  :OUTPUT;
)
BEGIN
        TABLE
        m/io, addr[15..0]    => rom, ram, print, sp[ ];
        1, B"00xxxxxxxxxxxxxx" => 1, 0, 0, B"00";
        1, B"10xxxxxxxxxxxxxx" => 0, 1, 0, B"00";
```

```
              0, B"000000101000000000" => 0, 0, 1, B"00";
              0, B"000011010000010000" => 0, 0, 0, B"01";
              END TABLE;
     END;
```

Instead of specifying all possible combinations, we can use x for "don't care" to indicate that output does not depend on the input corresponding to that position of x.

4.2.7 Implementing Active-Low Logic

An active-low control signal becomes active when its value is GND. An example of the circuit that uses active-low control signals is given in Example 4.10. The circuit performs daisy-chain arbitration. This module requests the bus access of the preceding module in the daisy chain. It receives requests for bus access from itself and from the next module in the chain. Bus access is granted to the highest-priority module that requests it.

Example 4.11 Active Low Daisy-Chain Arbitrator.

```
     SUBDESIGN daisy-chain
     (
             /local_request      : INPUT;
             /local_grant        : OUTPUT;
             /request_in         : INPUT; %from lower prior%
             /request_out        : OUTPUT;%to higher prior%
             /grant_in           : INPUT; %from higher prior%
             /grant_out          : OUTPUT; %to lower prior%
     )
     BEGIN
     DEFAULTS
             /local_grant = Vcc; %active-low output%
             /request_out = Vcc; %signals should%
             /grant_out = Vcc;   %default to Vcc%
     END DEFAULTS;

     IF /request_in == GND # /local_request == GND THEN
             /request_out = GND;
     END IF;
     IF /grant_in == GND THEN
             IF /local_request == GND THEN
```

```
                        /local_grant = GND;
                ELSIF /request_in == GND THEN
                        /grant_out = GND;
                END IF;
        END IF;
        END;
```

All signals in Example 4.11 are active-low. It is recommended to indicate that the signal is active low by using some indication, in this case a slash (/), as part of the signal name. The Defaults Statements in the example specify that a signal is assigned to Vcc when it is not active.

4.2.8 Implementing Bidirectional Pins

AHDL allows I/O pins in Altera devices to be configured as bidirectional pins. Bidirectional pins can be specified with a BIDIR port that is connected to the output of a TRI primitive. The signal between the pin and tri-state buffer is a bidirectional signal that can be used to drive other logic in the project.

Example 4.12 below shows an implementation of a register that samples the value found on a tri-state bus. It can also drive the stored value back to the bus. The bidirectional I/O signal, driven by TRI, is used as the D input to a D flip-flop (DFF). Commas are used to separate inputs to the D flip-flop, including placeholders for the CLRN and PRN signals that default to the inactive state.

Example 4.12 Bus Register.

```
        SUBDESIGN bus_reg
        (
                clk    : INPUT;
                oe     : INPUT;
                io     : BIDIR;
        )
        BEGIN
                io = TRI(DFF(io, clk, , ), oe);
        END;
```

A GDF equivalent to bus_reg example is shown in Figure 4.5.

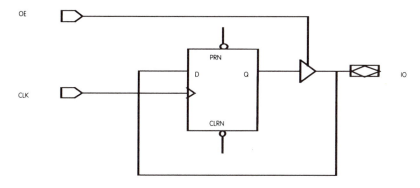

Figure 4.5 Example of a Bidirectional Pin

It is also possible to connect a bidirectional pin from a lower-level TDF to a top-level pin. The bidirectional port of the macrofunction should be assigned to a bidirectional pin. Example 4.13 shows the use of four instances of the bus_reg macrofunction

Example 4.13 Bidirectional 4-Bit Port.

```
TITLE "bidirectional 4-bit port";

FUNCTION bus_reg (clk, oe) RETURNS (io);

SUBDESIGN bidir
(
        clk, oe        : INPUT;
        io[3..0]       : BIDIR;
)
BEGIN
        io0 = bus_reg(clk, oe);
        io1 = bus_reg(clk, oe);
        io2 = bus_reg(clk, oe);
        io3 = bus_reg(clk, oe);
END;
```

The instances of bus_reg are used in-line in the corresponding AHDL statements.

4.3 Designing Sequential logic

Sequential logic is usually implemented in AHDL with state machines, registers, or latches and must include one or more flip-flops.

4.3.1 Declaring Registers and Registered Outputs

Registers are used to store data values, hold count values, and synchronize data with a Clock signal. Registers can be declared with a register declaration in the Variable Section. A port of an instance can be used to connect an instance of a primitive, macrofunction, or state machine to other logic in a TDF. A port of an instance uses the format:

> Instance_name.Port_name

The Port_name is an input or output of a primitive, macrofunction, or state machine, and is synonymous with a pin name in the GDF. Example 4.14 contains a byte register that latches values of the d inputs onto the q outputs on the rising edge of the Clock when the load input is high.

Example 4.14 Byte Register Design.

```
SUBDESIGN register
(
        clk, load, d[7..0]              : INPUT;
        q[7..0]                         : OUTPUT;
)
VARIABLE
        ff[7..0]                        : DFFE;
BEGIN
        ff[ ].clk = clk;
        ff[ ].ana = load;
        ff[ ].d = d[ ];
        q[ ] = ff[ ].q;
END;
```

All four statements in the Logic Section of the subdesign are evaluated concurrently. Instead of D flip-flops, other types of flip-flops can be declared in the Variable Section. A GDF equivalent to the above TDF is shown in Figure 4.6

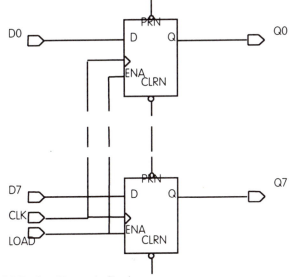

Figure 4.6 GDF of an Example Register

Registered outputs of a subdesign can be declared as D flip-flops in the Variable Section. Example 4.15 is similar to the previous one, but has registered outputs.

Example 4.15 Registered Output Byte Register.

```
SUBDESIGN reg_out
(
        clk, load, d[7..0]    :INPUT;
        q[7..0]               :OUTPUT;
)
VARIABLE
        q[7..0]               :DFFE; %also decl.as outp%
BEGIN
        q[].clk = clk;
        q[].ena = load;
        q[] = d[];
END;
```

Each Enable D flip-flop declared in the Variable Section feeds an output with the same name, so it is possible to refer to the q outputs of the declared flip-flops without using the q port of the flip-flops. The register's output does not change until the rising edge of the Clock. The Clock of the register is defined using

<output_pin_name>.clk

for the register input in the Logic Section. A global Clock can be defined with the GLOBAL primitive.

4.3.2 Creating Counters

Counters use sequential logic to count Clock or other pulses. Counters are usually defined with D flip-flops and IF Statements. Example 4.15 shows a 16-bit loadable up counter that can be cleared to zero.

Example 4.16 16-Bit Loadable Up Counter.

```
SUBDESIGN cnt
(
        clk, load, ena, clr, d[15..0]      :INPUT;
        q[15..0]                           :OUTPUT;
)
VARIABLE
        count[15..0]                       :DFF;
BEGIN
        count[].clk = clk;
        count[].clrn = !clr;

        IF load THEN
                count[].d = d[];
        ELSIF ena THEN
                count[].d = count[].q + 1;
        ELSE
                count[].d = count[].q;
        END IF;

        q[] = count[];
```

```
END;
```

In this example, 16 D flip-flops are declared in the Variable Section and assigned the names count0 through count15. The IF Statement determines the value that is loaded into the flip-flops on the rising Clock edge, depending on the value of control variables.

4.3.3 State Machines

AHDL enables easy implementation of state machines. The language is structured so that designers can either assign state bits by themselves, or allow the Compiler to do the work. If the task is performed by the Compiler, state assignment is done by minimizing the required logic resources. The designer only has to draw the state diagram and construct a next-state table. The Compiler then performs the following functions automatically:

- Assigns bits, select a T or D flip-flops to the bits

- Assigns state values

- Applies logic synthesis techniques to derive the excitation equations

The designer is allowed to specify state machine transitions in a TDF using Truth Table Statement as well. In that case, the following items must be included in the TDF:

- State Machine Declaration (Variable Section)

- Boolean Control Equations (Logic Section)

- State Transitions (Logic Section)

AHDL machines can be exported or imported between TDFs and GDFs, or TDFs and WDFs by specifying an input or output signal as a machine port in the Subdesign Section.

A state machine can be created by declaring the name of the state machine, its states, and optionally the state machine bits in the State Machine Declaration of the Variable Section. Example 4.17 represents the state machine with the functionality of a D flip-flop.

Example 4.17 State Machine D Flip-Flop.

```
SUBDESIGN dflip_flop
(
        clk, reset, d : INPUT;
        q                        : OUTPUT;
)
VARIABLE
        ss: MACHINE WITH STATES (s0, s1);
BEGIN
        ss.clk = clk;
        ss.reset = reset;

        CASE ss IS
                WHEN s0 =>
                        q = GND;

                        IF d THEN
                                ss = s1;
                        END IF;
                WHEN s1 =>
                        q = Vcc;

                        IF !d THEN
                                ss = s0;
                        END IF;
                END CASE;
END;
```

The states of a machine are defined as s0 and s1, and no state bits are declared. The GDF equivalent to this state machine is shown in Figure 4.7. A number of signals are used to control the flip-flops in the state machine. In a more general case, the states are represented by a number of flip-flops that form a state register. In Example 4.17, external clock and reset signals control directly clk and reset inputs of the state machine flip-flop. Obviously, the expression that specifies creation of these signals can be any Boolean expressions.

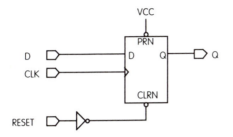

Figure 4.7 GDF Equivalent to the State Machine from Example

Clock, Reset, and Clock enable signals control the flip-flops of the state register. These signals are specified with Boolean equations in the Logic Section.

In Example 4.17, the state machine Clock is driven by the input clk. The state machine's asynchronous Reset signal is driven by reset, which is active high. To connect the Clock Enable signal in the TDF, we would add the line

```
ena     : INPUT;
```

to the Subdesign Section and the Boolean equation,

```
ss.ena = ena;
```

to the Logic section.

An output value can be defined with an IF or Case Statement. In our example, output q is assigned to GND when state machine ss is in state s0, and to value Vcc when the machine is in state s1. These assignments are made in WHEN clauses of the Case Statement. Output values can also be defined in truth tables.

State machine transitions define the conditions under which the state machine changes to a new state. The states must be assigned within a single behavioral construct to specify state machine transitions. For this purpose, it is recommended to use Case or Truth Table Statements. The transitions out of each state are defined in WHEN clauses of the Case Statement.

The state bits, which represent outputs of the flip-flops used by a state machine, are usually assigned by the MAX+PLUS II Compiler. However, the designer is allowed to make these assignments explicitly in the State Machine declaration. An example of such an assignment is shown in Example 4.18.

Example 4.18 Direct State Bit Assignment.

```
SUBDESIGN stat_assign
(
        clk, reset, ccw, cw :INPUT;
        phase[3..0]                     :OUTPUT;
)
VARIABLE
        ss: STATE MACHINE OF BITS (phase[3..0])
                WITH STATES (
                        s0 = B"0001",
                        s1 = B"0010",
                        s2 = B"0100",
                        s3 = B"1000");
BEGIN
        ss.clk = clk;
        ss.reset = reset;

TABLE
        ss,     ccw, cw =>   ss;

        s0,  1,    x   =>  s3;
        s0,  x,    1   =>  s1;
        s1,  1,    x   =>  s0;
        s1,  x,    1   =>  s2;
        s2,  1,    x   =>  s1;
        s2,  x,    1   =>  s3;
        s3,  1,    x   =>  s2;
        s3,  x,    1   =>  s0;
END TABLE;
END;
```

In Example 4.18, the phase[3..0] outputs declared in the Subdesign Section are also declared as bits of the state machine ss.

4.3.4 State Machines with Synchronous Outputs

State machines where the present state depends only on its previous inputs and previous state, and the present output depends only on the present state, are called Moore State Machines. Outputs of Moore State Machines can be specified in the WITH STATES clause of the State Machine Declaration. The Example 4.19 implements the Moore State Machine.

Example 4.19 Moore State Machine.

```
SUBDESIGN moore
(
        clk, reset, y : INPUT;
        z                     : OUTPUT;
)

VARIABLE

                        % current              current%
                        % state                output%
ss: MACHINE OF BITS          (z)
     WITH STATES (s0 =       0,
                     s1 =   1,
                     s2 =   1,
                     s3 =   0);
BEGIN
        ss.clk = clk;
        ss.reset = reset;

        TABLE
        % current        current         next %
        % state          input           state %
           ss,  y         =>       ss;

           s0,  0         =>       s0;
           s0,  1         =>       s2;
           s1,  0         =>       s0;
           s1,  1         =>       s2;
           s2,  0         =>       s2;
           s2,  1         =>       s3;
           s3,  0         =>       s3;
           s3,  1         =>       s1;
        END TABLE;
```

```
END;
```

The state machine is defined with a State machine declaration. The state transitions are defined in a next-state table, which is implemented with a Truth Table Statement. In this example, machine ss has four states and only one state bit z. The Compiler automatically adds another bit and makes appropriate assignments to produce a four- state machine. When state values are used as outputs, as in example above, the project may use fewer logic cells, but the logic cells may require more logic to drive their flip-flop inputs.

Another way to design state machines with synchronous outputs is to omit state value assignments and to explicitly declare output flip-flops. This method is illustrated in Example 4.20.

Example 4.20 Moore Machine with Explicit Output D Flip-Flops.

```
SUBDESIGN moore
(
        clk, reset, y : INPUT;
        z                      : OUTPUT;
)
VARIABLE
        ss: MACHINE WITH STATES (s0,s1,s2,s3);
        zd : NODE;

BEGIN
        ss.clk = clk;
        ss.reset = reset;

        z = DFF(zd, clk, Vcc, Vcc);

        TABLE
        % current      current        next       next %
        % state        input          state      output %
          ss,          y        =>    ss,             zd;
          s0,          0        =>    s0,             0;
          s0,          1        =>    s2,             1;
          s1,          0        =>    s0,             0;
          s1,          1        =>    s2,             1;
          s2,          0        =>    s2,             1;
          s2,          1        =>    s3,             0;
          s3,          0        =>    s3,             0;
          s3,          1        =>    s1,             1;
```

```
              END TABLE;
       END;
```

This example includes a "next output" column after the "next state" column in the Truth Table Statement. This method uses a D flip-flop, called with an in-line reference, to synchronize the outputs with the Clock.

4.3.5 State Machines with Asynchronous Outputs

State machines where the outputs are a function of the current inputs and the current states are called Mealy State Machines. AHDL supports implementation of state machines with asynchronous outputs. The outputs of Mealy State Machines may change when inputs change, regardless of Clock transitions. Example 4.21 shows a state machine with asynchronous outputs.

Example 4.21 State Machine with Asynchronous Outputs.

```
       SUBDESIGN mealy
       (
              clk, reset, y : INPUT;
              z                    : OUTPUT;
       )
       VARIABLE
              ss: MACHINE WITH STATES (s0, s1, s2, s3);

       BEGIN
              ss.clk = clk;
              ss.reset = reset;

              TABLE
              % current      current        current    next%
              % state        input          output     state%
                ss,          y      =>       z,          ss;
                s0,          0      =>       0,          s0;
                s0,          1      =>       1,          s1;
                s1,          0      =>       0,          s1;
                s1,          1      =>       1,          s2;
                s2,          0      =>       0,          s2;
                s2,          1      =>       1,          s3;
                s3,          0      =>       0,          s3;
                s3,          1      =>       1,          s0;
```

```
        END TABLE;
END;
```

4.4 Implementing a Hierarchical Project

AHDL TDFs can be mixed with GDFs, WDFs, EDIF Input Files, Altera Design Files, State Machine Files, and other TDFs in a project hierarchy. Lower level files in a project hierarchy can either be Altera-provided macrofunctions or user defined (custom) macrofunctions.

4.4.1 Using Altera-provided Macrofunctions

MAX+PLUS II includes a large library of standard 74-series, bus, architecture optimized, and application-specific macrofunctions which can be used to create a hierarchical logic design. These macrofunctions are installed in the \maxplus2\max2lib directory and its subdirectories.

There are two ways to call (insert an instance) a macrofunction in AHDL. One way is to declare a variable of type <macrofunction> in an Instance Declaration in the Variable Section and use ports of the instance of the macrofunction in the Logic Section. In this method, the names of the ports are important. The second way is to use a macrofunction reference in the Logic section of the TDF. In this method, the order of the ports is important.

The inputs and outputs of macrofunctions are listed in the Function Prototype Statement. A Function Prototype Statement can also be saved in an Include File and imported into a TDF with an Include Statement. Include Files for all macrofunctions are provided in the \maxplus2\max2inc directory.

Example 4.22 shows the connection of a 4-bit counter to a 4-to-16 decoder. Macrofunctions are called with Instance Declarations in the Variable Section.

Example 4.22 4-Bit Counter to 4-16 Decoder Connection.

```
        INCLUDE "4count";
        INCLUDE "16dmux";
```

```
SUBDESIGN macro1
(
        clk            : INPUT;
        out[15..0]   :OUTPUT;
)

VARIABLE
        counter        : 4count;
        decoder        : 16dmux;

BEGIN
        counter.clk = clk;
        counter.dnup = GND;
        decoder.(d,c,b,a) = counter.(qd,qc,qb,qa);
        out[15..0] = decoder.q[15..0];
END;
```

Include Statements are used to Import Function Prototypes for the two Altera provided macrofunctions. In the Variable Section, the variable counter is declared as an instance of the 4count macrofunction, and the variable decoder is declared as an instance of the 16dmux macrofunction. The input ports for both macrofunctions, in the format <Instance_name>.Port_name, are defined on the left side of the Boolean equations in the Logic Section; the output ports are defined on the right. The order of the ports in the Function Prototypes is not important because the port names are explicitly listed in the Logic Section. A GDF that is equivalent to the example above is shown in Figure 4.8.

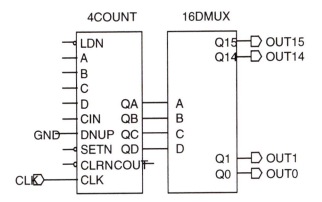

Figure 4.8 GDF Equivalent to Example TDF

The same functionality as in Example 4.22 can be implemented using in line references as shown in Example 4.23.

Example 4.23 In Line References for Counter and Decoder Connections.

```
INCLUDE "4count";
INCLUDE "16dmux";

SUBDESIGN macro2
(
        clk             :INPUT;
        out[15..0]      :OUTPUT;
)
VARIABLE
        q[3..0]         :NODE;
BEGIN
        (q[3..0], )= 4count (clk,,,,, GND,,,,);
        out[15..0] = 16dmux (q[3..0]);
END;
```

The in-line reference for the functions 4count and 16dmux appear on the right side of the Boolean equations in the Logic Section. The Function Prototypes for the two macrofunctions, which are stored in the Include Files 4count.inc and 16dmux.inc, are shown in Example 4.24.

Example 4.24 Function Prototypes for In Line References.

```
FUNCTION 4count (clk,clrn,setn,ldn,cin,dnup,d,c,b,a)
        RETURNS (qd, qc, qb, qa, cout);

FUNCTION 16dmux (d, c, b, a)
        RETURNS (q[15..0]);
```

The order of ports is important because there is a one-to-one correspondence between the order of the ports in the Function Prototype and the ports defined in the Logic Section. In Example 4.24, commas are used as placeholders for ports that are not explicitly connected.

4.4.2 Creating and Using Custom Macrofunctions

Custom macrofunctions can be easily created and used in AHDL by performing the following tasks:

- Create the logic for the macrofunction in a design file.

- Specify the macrofunction's ports with a Function Prototype Statement. This, in turn, provides a shorthand description of a function, listing the name and its input, output, and bidirectional ports. Machine ports can also be used for macrofunctions that import or export state machines. The Function Prototype Statement can also be placed in an Include File and called with an Include Statement in the file.

- Insert an instance of the macrofunction with an Instance Declaration or an in-line reference.

- Use the macrofunction in the file.

To use a macrofunction, a Function Prototype must be included in the current TDF, or in the Include Statement used to include the information from an Include File's Function Prototype. Example 4.25 shows the implementation of a keyboard encoder for a 16-key calculator keyboard.

Example 4.25 Keyboard Encoder.

```
TITLE "Keyboard encoder"

INCLUDE "74151";
INCLUDE "74154";
INCLUDE "4count";

FUNCTION debounce (clk, key_pressed);
       RETURNS (pulse);

SUBDESIGN keyboard
(
       clk           : INPUT; % 50 KHz clock %
       col[3..0]     : INPUT; % keyboard columns %
       row[3..0], d[3..0] : OUTPUT; % k-rows,key code%
       strobe : OUTPUT; % key code is valid %
```

```
        )
        VARIABLE
                key_pressed : NODE; % Vcc when key d[3..0] %
                                         % is pressed %
                mux          : 74151;
                decoder      : 74154;
                counter      : 4count;
                opencol[3..0] : TRI;
        BEGIN
                % drive keyb.rows with a decoder and open %
                % collector outputs %
                row[ ] = opencol[ ].out;
                opencol[ ].in = GND;
                opencol[ ].oe = decoder.(o0n,o1n,o2n,o3n);
                decoder.(b,a) = counter.(qd,qc);

                % sense keyb.columns with a multiplexer %
                mux.d[3..0] = col[3..0];
                mux.(b,a) = counter.(qb,qa);
                key_pressed = !mux.y;

                % scan keyb. until a key is pressed &
                % drive key's code onto d[ ] outputs %

                counter.clk = clk;
                counter.cin = !key_pressed;
                d[ ] = counter.(qd,qc,qb,qa);

                % generate strobe when key has settled %

                strobe = debounce(clk, key_pressed);
        END
```

Include Statements include Function Prototypes for the Altera provided macrofunctions 4count, 74151, and 74154. A separate Function Prototype Statement specifies the ports of the custom macrofunction debounce within the TDF rather than in an Include File. Instances of Altera provided macrofunctions are called with Instance Declarations in the Variable Section; an instance of the debounce macrofunction is called with an in-line reference in the Logic Section.

4.4.3 Using Parametrized Functions

A number of MAX+PLUS II functions are parametrized, allowing declaration of the parameter value at the compilation time. For example, parameters are used to specify the width of ports that represent input operands and results of operation of a functional unit. Parametrized functions are instantiated with an in-line logic function reference or an Instance Declaration in the same way as unparametrized functions. However, a few additional steps are required to declare the values of parameters, providing essentially customization of the function to the designer's requirements. These steps include:

- Use of the WITH clause, that lists parameters used by the instance. If parameter values are not supplied within the instance, they must be provided somewhere else within the project.
- Specification of the values of unconnected pins. This requirement comes from the fact that the parametrized functions do not have default values for unconnected inputs.

The inputs, outputs, and parameters of the function are declared with Function Prototype Statement, or they can be provided from corresponding Include Files. For example, if we want to use MAX+PLUS II multiplier library parametrized module (LPM) lpm_mult, its Function Prototype is shown in Example 4.26

Example 4.26 Function Prototype of the Multiplier LPM

```
FUNCTION lpm_mult (dataa[(LPM_WIDTHA-1)..0], datab[(LPM_WIDTHB-
1)..0], sum[(LPM_WIDTHS-1)..0], aclr, clock)
   WITH (LPM_WIDTHA, LPM_WIDTHB, LPM_WIDTHP, LPM_WIDTHS,
LPM_REPRESENTATION, LPM_PIPELINE, LATENCY, INPUT_A_IS_CONSTANT,
INPUT_B_IS_CONSTANT, USE_EAB)
   RETURNS (result[LPM_WIDTHP-1..0]);
```

This function provides multiplication of input operands a and b and addition of the partial sum to provide the output result. Clock and aclr control signals are used for pipelined operation of the multiplier to clock and clear intermediate registers.

A number of parameters using WITH clause are provided including those to specify the widths and nature of all operands and result (variable and constants), as well as the use of FPLD resources (implementation using EABs in the case of

FLEX 10K devices). Only widths of inputs and result are required, while the other parameters are optional. The design presented in Example 4.27 shows the use of lpm_mult function:

Example 4.27 Use of lpm_mult function

```
INCLUDE "lpm_mult.inc";

SUBDESIGN mult8x8
(
        a[7..0], b[7..0]     : INPUT;
        c[15..0]             : OUTPUT;
)

BEGIN

c[] = lpm_mult(a[], b[],0, , )
        WITH (lpm_widtha=8, lpm_widthb=8,
        lpm_widths=1, lpm_widthp=16);
END;
```

It should be noted that the width must be a positive number. Also, placeholders are used instead of clock and clear inputs.

Another possibility is to use Instance Declaration of the multiplier, as in Example 4.28

Example 4.28 Use of instance declaration of the Multiplier

```
INCLUDE "lpm_mult.inc";

SUBDESIGN mult8x8
(
        a[7..0], b[7..0]     : INPUT;
        c[15..0]             : OUTPUT;
)
VARIABLE
        8x8mult: lpm_mult WITH(lpm_widtha=8,
        lpm_widthb=8, lpm_widths=1, lpm_widthr=16);
BEGIN
        8x8mult.dataa[]= a[];
```

```
8x8mult.datab[]= b[];
c[]= 8x8mult.result[];
```

END;

LPMs currently available in MAX+PLUS II library are listed in Section 4.7.4.

4.4.4 Implementing RAM and ROM

Besides providing standard combinational and sequential parametrized logic modules, MAX+PLUS II provides a number of very useful memory functions that can be a part of user's designs. Their implementation is not equally efficient in all Altera's devices, but higher capacity is obviously achieved in FLEX 10K devices due to existence of embedded array blocks.

The following memory megafunctions can be used to implement RAM and ROM:

- lpm_ram_dq   Synchronous or asynchronous memory with separate input and output ports
- lpm_ram_io   Synchronous or asynchronous memory with single I/O port
- lpm_rom   Synchronous or asynchronous read-only memory
- csdpram   Cycle-shared dual port memory
- csfifo   Cycle-shared first-in first-out (FIFO) buffer

Parameters are used to determine the input and output data widths, the number of data words stored in memory, whether data inputs, address and control inputs and outputs are registered or not, whether an initial memory content file is to be included, etc.

For example, synchronous or asynchronous read-only (ROM) memory megafunction is represented by the prototype function, as in Example 4.29

Example 4.29 Synchronous or Asynchronous ROM Megafunction

```
FUNCTION   lpm_rom   (address[LPM_WIDTHAD-1..0],   inclock,
outclock, memenab)
    WITH      (LPM_WIDTH,      LPM_WIDTHAD,      LPM_NUMWORDS,
LPM_FILE, LPM_ADDRESS_CONTROL, LPM_OUTDATA)
    RETURNS (q[LPM_WIDTH-1..0]);
```

Input ports are address lines with the number of lines specified by lpm_widthad parameter, inclock and outclock that specify frequency for input and output registers, and memory enable, while the output port is presented by a number of data lines given with parameter lpm_width. The ROM megafunction can be used either by in-line reference or by Instnce Declaration, as it was shown in Example 4.29.

4.5 Reserved Keywords and Symbols

Reserved keywords are used for beginnings, endings, and transitions of AHDL statements and as predefined constant values such as GND and Vcc. Table 4.1 shows all AHDL reserved keywords in alphabetical order.

Table 4.1 AHDL reserved keywords

AND	BEGIN	BIDIR	BITS	BURIED
CARRY	CASCADE	CASE	CLIQUE	CONNECTED_PINS
CONSTANT	DEFAULTS	DEFINE	DEVICE	DFF
DFFE	DIV	ELSE	ELSIF	END
EXP	FOR	FUNCTION	GENERATE	GLOBAL
GND	HELP_ID	IF	INCLUDE	INPUT
IS	JKFF	JKFFE	LATCH	LCELL
LOG2	MACHINE	MEMORY	MOD	MCELL
NAND	NODE	NOR	NOT	OF
OPNDRN	OPTIONS	OR	OTHERS	OUTPUT
PARAMETERS	REPORT	RETURNS	SEGMENTS	SEVERITY
SOFT	SRFF	SRFFE	STATES	SUBDESIGN
TABLE	TFF	TFFE	THEN	TITLE
TO	TRI	TRI_STATE_NODE	VARIABLE	VCC
WHEN	WIRE	WITH	X	XNOR
XOR				

Some symbols having predefined meanings in AHDL are described in Table 4.2.

Table 4.2 AHDL symbols with predefined meanings

Symbol	Function
_ (underscore) - (dash)	User-defined identifiers used as legal characters in symbolic names
% (percent)	Encloses comments
( ) (left & right parentheses)	Enclose and define sequential group names. Enclose pin names in Subdesign Section and Function Prototype. Optionally enclose inputs and outputs of truth Tables in Truth Table Statements. Enclose states of State Machine Declarations. Enclose highest priority operations in Boolean Expressions. Enclose options in a Design Section (within Resource Assignment Statement).
[ ] (left and right brackets)	Enclose the number range of a decimal group name.
'...' (single quotation marks)	Enclose quoted symbolic names.
"..." (double quotation marks)	Enclose string in Title Statements. Enclose pathname in Include Statements. Enclose digits in non-decimal numbers. Optionally enclose design name and device name in Design Section. Optionally enclose clique name in Clique Assignment Section.
. (period)	Separates symbolic names of macrofunction or primitive variables from port names. Separates extensions from filenames.
.. (elipsis)	Separates most significant bit (MSB) from least Significant bit (LSB) in ranges.
; (semicolon)	Ends AHDL statements and sections.
, (comma)	Separates members of sequential groups and lists.
: (colon)	Separates symbolic names from types in declarations and resource assignments.
@ (at)	Assigns symbolic nodes to device pins and logic cells in Resource Assignments Statements.

=> (arrow)	Separates inputs from outputs in Truth Table Statements. Separates WHEN clauses from Boolean expressions in Case Statements.
= (equals)	Assigns default GND and VCC values to inputs in Subdesign Section. Assigns settings to options. Assigns values to state machine states. Assigns values in Boolean equations.

AHDL supports three types of names, Symbolic names, Subdesign names, and Port names.

Symbolic names are user-defined identifiers. They are used to name internal and external nodes, constants, state machine variables, state bits, states, and instances.

Subdesign names are user-defined names for lower-level design files; they must be the same as the TDF filename.

Port names are symbolic names that identify input or output of a primitive or macrofunction.

Names can be used in quoted or unquoted notation. Quoted names are enclosed in single quotation marks. Quotes are not included in pinstub names that are shown in the symbol for a TDF.

4.6 Boolean Expressions

The result of every Boolean expression must be the same width as the node or group (on the left side of an equation) to which it is eventually assigned. The logical operators for Boolean expressions are shown in Table 4.3.

Table 4.3 Logical Operators for Boolean expressions.

Operator	Description	Example
! (NOT)	one's complement (prefix inverter)	!alpha NOT alpha
& (AND)	AND	Alpha & beta alpha AND beta
!& (NAND)	AND inverter	A[3..0]!&b [5..2] a[3..0]NAND b[5..2]
# (OR)	OR	All # all1 all OR all1
!# (NOR)	OR inverter	A[3..1]!# b[4..2] a[3..1]NOR b[4..2]
$ (XOR)	Exclusive OR	Ami $ bami ami XOR bami
!$ (XNOR)	Exclusive NOR	X1 !$ X2 x1 XNOR x2

Each operator represents a 2-input logic gate, except the NOT operator which is a prefix inverter. Expressions using these operators are interpreted differently depending on whether the operands are single nodes, groups, or numbers.

Three operand types are possible with the NOT operator. If the operand is a single node, GND, or Vcc, a single inversion operation is performed. If the operand is a group of nodes, every member of the group passes through an inverter. If the operand is a number, it is treated as a binary number with as many bits as the group context in which it is used and every bit is inverted. For example: !5 in a three-member group is interpreted as !B"101" = B"010".

Five operand combinations are possible with the binary operators and each of these combinations is interpreted differently:

• If both operands are single nodes or the constants GND or Vcc, the operator performs the logical operation on two elements.

- If both operands are groups of nodes, the operator produces a bit wise set of operations between the groups. The groups must be of the same size.

- If one operand is a single node (GND or Vcc) and the other operand is a group of nodes, the single node or constant is duplicated to form a group of the same size as the other operand. The expression is then treated as group operation.

- If both operands are numbers, the shorter number is sign extended to match the size of the other number. The expression is then treated as a group operation.

- If one operand is a number and the other is a node or group of nodes, the number is truncated or sign extended to match the size of the group.

Arithmetic operators are used to perform arithmetic addition and subtraction operations on groups and numbers. Table 4.4 shows the arithmetic operators in AHDL.

Table 4.4 AHDL arithmetic operators.

Operator	Description	Example
+ (unary)	Positive	+5
- (unary)	Negative	-b[3..0]
+	Addition	a[3..0] + b[3..0]
-	Subtraction	a_1[ ] - b_1[ ]

The + unary operator does not effect the operand. The - unary operator interprets its operand as a binary representation of a number. It then performs a two's complement unary minus operation on the operand. In the case of arithmetic operators the following rules apply:

- The operands must be groups of nodes or numbers.

- If both operands are groups of nodes, the groups must be of the same size.

- If both operands are numbers, the shorter operand is sign-extended to match the size of the other operand.

- If one operand is a number and the other is a group of nodes, the number is truncated or sign extended to match the size of the group. In the case of truncation of any significant bits, the compiler generates an error message.

Comparators are used to compare single nodes or groups. There are two types of comparators: logical and arithmetic. All types of comparators in AHDL are presented in Table 4.5.

Table 4.5 AHDL Comparators.

Comparator	Type	Description	Example
==	Logical	equal to	addr[2..0] == B"101"
!=	Logical	not equal to	a_1 != a_2
<	Arithmetic	less than	a[ ] < b[ ]
<=	Arithmetic	less than or equal	c[ ] <= d[ ]
>	Arithmetic	Greater than	a[ ] > b[ ]
>=	Arithmetic	Greater than or equal	c[ ] >= d[ ]

The logical equal to operator (==) is used exclusively in Boolean expressions. Logical comparators can compare single nodes, groups (of the same size) or numbers. Comparison is performed on a bit wise basis and returns Vcc when the comparison is true, and GND when the comparison is false. Arithmetic comparators can only compare groups of nodes or numbers. Each group is interpreted as a positive binary number and compared to the other group.

Priority of evaluation of logical and arithmetic operators and comparators is given in Table 4.6 (operations of equal priority are evaluated from left to right with the possibility to change the order using parentheses).

Table 4.6 AHDL Comparator and Operator Priority.

Priority	Operator/Comparator
1	– (negative)
1	! (NOT)
2	+ (addition)
2	– (subtraction)
3	== (equal to)
3	!= (not equal to)
3	< (less than)
3	<= (less than or equal to)
3	> (greater than)
3	>= (greater than or equal to)
4	& (AND)
4	!& (NAND)
5	$ (XOR)
5	!$ (XNOR)
6	# (OR)
6	!# (NOR)

4.7 Library Functions

AHDL TDFs use statements, operators, and keywords to replace some GDF primitives. Function Prototypes for these primitives are not required in TDFs. However, they can be used to redefine the calling order of the primitive inputs.

4.7.1 Buffer Primitives

Buffer primitives allow control of the logic synthesis process. In most circumstances it is advisable to let the compiler indicate when and where to insert the buffers in order to support logic expansion.

1. CARRY

Function Prototype: FUNCTION carry (in)

```
                          RETURNS (out);
```

The carry buffer designates the carry out logic for a function and acts as the carry in to another function. It is supported only by the FLEX 8000 family of devices.

2. CASCADE

Function Prototype: FUNCTION cascade (in)
                           RETURNS (out);

The cascade buffer designates the cascade out function from an AND or an OR gate, and acts as an cascade-in to another AND or OR gate. It is supported only by the FLEX 8000 family of devices.

3. EXP

Function Prototype: FUNCTION EXP(in)
                           RETURNS (out);

The EXP expander buffer specifies that an expander product term is desired in the project. The expander product is inverted in the device. This feature is supported only for MAX devices. In other families it is treated as a NOT gate.

4. GLOBAL

Function Prototype: FUNCTION GLOBAL (in)
                           RETURNS (out);

The global buffer indicates that a signal must use a global (synchronous) Clock, Clear, Preset, or Output Enable signal, instead of signals generated by internal logic or driven by ordinary I/O pins. If an input port feeds directly to the input of GLOBAL, the output of GLOBAL can be used to feed a Clock, Clear, Preset, or Output Enable to a primitive. A direct connection must exist from the output of GLOBAL to the input of a register or a TRI buffer. A NOT gate may be required between the input pin and GLOBAL when the GLOBAL buffer feeds the Output Enable of a TRI buffer.

Global signals propagate more quickly than array signals, and are used to implement global clocking in a portion or all of the project. Legal and illegal use of GLOBAL buffer are shown in Figure 4.9.

Legal uses of GLOBAL buffer

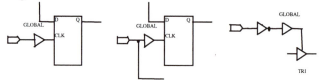

Illegal uses of GLOBAL buffer

Figure 4.9 Legal and Illegal Use of the GLOBAL Buffer

5. LCELL

Function Prototype:
```
FUNCTION lcell (in)
    RETURNS (out);
```

The LCELL buffer allocates a logic cell for the project. It produces the true and complement of a logic function and makes both available to all logic in the device. An LCELL always consumes one logic cell

6. OPNDRN

Function Prototype:
```
FUNCTION opndrn (in)
    returns (out);
```

The OPNDRN primitive is equivalent to a TRI primitive whose Output Enable input can be any signal, but whose primary input is fed by GND primitive. If the input to the OPNDRN primitive is low, the output will be low. If the input is high, the output will be a high-impedance logic level.

The OPNDRN primitive is supported only for FLEX 10K device family. It is automatically converted to a TRI primitive for other devices.

7. SOFT

Function Prototype: FUNCTION soft (in)
                           RETURNS (out);

The SOFT buffer specifies that a logic cell may be needed at a specific location in the project. The Logic Synthesizer examines whether a logic cell is needed. If it is needed, the SOFT is converted into an LCELL; if not, the SOFT buffer is removed. If the Compiler indicates that the project is too complex, a SOFT buffer can be inserted to prevent logic expansion. For example, a SOFT buffer can be added at the combinational output of a macrofunction to decouple two combinational circuits.

8. TRI

Function Prototype: FUNCTION TRI (in, oe)
                           RETURNS (out);

The TRI is a tri-state buffer with an input, output, and Output Enable (oe) signal. If the oe signal is high, the output is driven by input. If oe is low, the output is placed into a high-impedance state that allows the I/O pin to be used as an input pin. The oe defaults to Vcc. If oe of a TRI buffer is connected to Vcc or a logic function that will minimize to true, a TRI buffer can be converted into a SOFT buffer during logic synthesis. When using a TRI buffer, the following must be considered:

- A TRI buffer may only drive one BIDIR port. A BIDIR port must be used if feedback is included after the TRI buffer.

- If a TRI buffer feeds logic, it must also feed a BIDIR port. If it feeds a BIDIR port, it may not feed any other outputs.

- When oe is not tied to Vcc, the TRI buffer must feed an OUTPUT or BIDIR port. Internal signals may not be tri-stated.

4.7.2 Flip-flop and Latch Primitives

MAX+PLUS II flip-flop and latch primitives are listed together with their Function Prototypes in Table 4.7. All flip-flops are positive edge triggered and latches are level-sensitive. When the Latch or Clock Enable (ena) input is high, the flip-flop or latch passes the signal from that data input to q output. When the ena input is low, the state q is maintained regardless of the data input.

Table 4.7 Flip-flop and latch primitives.

Primitive	AHDL Function Prototype
LATCH	FUNCTION LATCH (d, ena) RETURNS (q);
DFF	FUNCTION DFF (d, clk, clrn, prn) RETURNS (q);
DFFE	FUNCTION DFFE (d, clk, clrn, prn, ena) RETURNS (q);
JKFF	FUNCTION JKFF (j, k, clk, clrn, prn) RETURNS (q);
JKFFE	FUNCTION JKFFE (j, k, clk, clrn, prn, ena) RETURNS (q);
SRFF	FUNCTION SRFF (s, r, clk, clrn, prn) RETURNS (q);
SRFFE	FUNCTION SRFFE (s, r, clk, clrn, prn, ena) RETURNS (q);
TFF	FUNCTION TFF (t, clk, clrn, prn) RETURNS (q);
TFFE	FUNCTION TFFE (t, clk, clrn, prn, ena) RETURNS (q);

Notes and definitions for Table 4.7:

clk	= Register Clock Input
clrn	= Clear Input
d, j, k, r, s, t	= Input from Logic Array
ena	= latch Enable or Clock Enable Input
prn	= Preset Input
q	= Output

4.7.3 Macrofunctions

MAX+PLUS II provides a number of standard macrofunctions that represent high level building blocks that may be used in logic design. The macrofunctions are automatically installed in the \maxplus2\max2lib directory and its subdirectories. The \maxplus2\max2inc directory contains an Include File with a Function Prototype for each macrofunction. All unused gates and flip-flops are automatically removed by the compiler. The input ports also have default signal values, so the unused inputs can simply be left unconnected. Most of the macrofunctions have the same names as their 74-series TTL equivalents, but some additional macrofunctions are also available. Refer to the relevant directories for the most recent list of available macrofunctions. Examples of macrofunctions are given in Table 4.8.

Table 4.8 Standard MAX+PLUS II Macrofunctions.

Macrofunction Type	Macrofunction Name	Description of Operation
Adder	8fadd 7480 74283	8-bit full adder Gated full adder 4-bit full adder with fast carry
Arithmetic Logic Unit	74181 74182	Arithmetic logic unit Look-ahead carry generator

Application specific	Pll	Rising- and falling-edge detector
Buffer	74240	Octal inverting 3-state buffer
	74541	Octal 3-state buffer
Comparator	8mcomp	8-bit magnitude comparator
	7485	4-bit magnitude comparator
	74688	8-bit identity comparator
Converter	74184	BCD-to-binary converter
Counter	gray4	Gray code counter
	7468	Dual decade counter
	7493	4-bit binary counter
	74191	4-bit up/down counter with asynch.
	74669	Load
		Synchr. 4-bit up/down counter
Decoder	16dmux	4-to-16 decoder
	7446	BCD-to-7-segment decoder
	74138	3-to-8 decoder
Digital Filter	74297	Digital phase-locked loop filter
EDAC	74630	16-bit parallel error detection &correction circuit
Encoder	74148	8-to-3 encoder
	74348	8-to-3 priority encoder with 3-state outputs
Frequency divider	7456	Frequency divider
Latch	Inpltch	Input latch
	7475	4-bit bistable latch
	74259	8-bit addressable latch with Clear
	74845	8-bit bus interface D latch with 3-state outputs
Multiplier	mult4	4-bit parallel multiplier
	74261	2-bit parallel binary multiplier
Multiplexer	21mux	2-to-1 multiplexer
	74151	8-to-1 multiplexer
	74157	Quad 2-to-1 multiplexer
	74356	8-to-1 data selector/multiplexer/register with 3-state outputs
Parity generator/ checker	74180	9-bit odd/even parity generator/checker
Rate multiplier	74167	Synchronous decade rate multiplier

Register	7470	AND-gated JK flip-flop with Preset and Clear
	7473	Dual JK flip-flop with Clear
	74171	Quad D flip-flops with Clear
	74173	4-bit D register
	74396	Octal storage register
Shift register	Barrelst	8-bit barrel shifter
	7491	Serial-in serial-out shift register
	7495	4-bit parallel-access shift register
	74198	8-bit bidirectional shift register
	74674	16-bit shift register
Storage register	7498	4-bit data selector/storage register
SSI Functions	Inhb	Inhibit gate
	7400	NAND2 gate
	7421	AND4 gate
	7432	OR2 gate
	74386	Quadruple XOR gate
True/ Complement I/O Element	7487	4-bit true/complement I/O element
	74265	Quadruple complementary output elements

4.7.4 Logic Paramtrized Modules

A list of LPMs supported by Altera for use in VHDL and other tools within MAX+PLUS II design environment is shown in the following table:

Table 4.9 List of LPMs

Name	Description
Gates	
Lpm_and	Multi-bit and gate for bit-wise and operation
Lpm_bustri	Multi-bit three-state buffer for unidirectional and bidirectional buffer Implementation
Lpm_clshift	Combinatorial Logic Shifter or Barrel Shifter
Lpm_constan	Constant Generator

t	
Lpm_decode	Decoder
Lpm_or	Multi-bit or gate for bit-wise or operation
Lpm_xor	Multi-bit xor gate for bit-wise xor operation
Lpm_inv	Multi-bit inverter
Busmux	Two-input multi-bit multiplexer (can be derived from lpm_mux)
Lpm_mux	Multi-input multi-bit multilexer
Mux	Single input multi-bit multiplexer (can be derived from lpm_mux)
Arithmetic Components	
Lpm_abs	Absolute value
Lpm_add_su b	Multi-bit Adder/Subtractor
Lpm_compar e	Two-input multi-bit comparator
Lpm_counter	Multi-bit counter with various control options
Lpm_mult	Multi-bit multiplier

4.7.5 Ports

A port is an input or output of a primitive, macrofunction, or state machine. A port can appear in three locations: the Subdesign Section, the Design Section, and the Logic Section.

A port that is an input or output of the current file is declared in the Subdesign Section. It appears in the following format:

Port_Name : Port_Type [=Default_Port_Value]

The following are port types available: INPUT, OUTPUT, BIDIR, MACHINE INPUT, and MACHINE OUTPUT.

When a TDF is the top-level file in hierarchy, the port name is synonymous with a pin name. The optional port value, GND or Vcc, can be specified for

INPUT and BIDIR port types. It is used only if the port is left unconnected when an instance of the TDF is used in a higher- level design file.

A port that is an input or output of the current file can be assigned to a pin, logic cell, or chip in the Design Section. Examples are given in earlier sections.

A port that is an input or output of an instance of a primitive or lower-level design file is used in the Logic Section. To connect a primitive, macrofunction, or state machine to other portions of a TDF, insert an instance of the primitive or macrofunction with an in-line reference or Instance Declaration, or declare the state machine with a State Machine Declaration. Then use ports of the function in the Logic Section. Port names are used in the following format:

Instance_name.Port_name

The Instance_name is a user defined name for a function. The Port_name is identical to the port name that is declared in the Logic Section of a lower level TDF or to a pin name in another type of design file.

All Altera-provided logic functions have predefined port names, which are shown in the Function Prototype. Commonly used names are shown in Table.

Table 4.10 List of Port Names

Port Name	Definition
.q	Output of a flip-flop or latch
.d	Data input to a D flip-flop or latch
.t	Toggle input to a T flip-flop
.j	J input to a JK flip-flop
.k	K input to a JK flip-flop
.s	Set input to a SR flip-flop
.r	Reset input to a SR flip-flop
.clk	Clock input to a flip-flop
.ena	Clock Enable input to a flip-flop, Latch Enable input to a latch, or Enable input to a state machine
.prn	Active-low Preset input to a flip-flop
.clrn	Active-low Clear input to a flip-flop
.reset	Active-high Reset input to a state machine
.oe	Output Enable input to a flip-flop

.in	Primary input to CARRY, CASCADE, EXP, GLOBGAL, LCELL, OPNDRN, SOFT, and TRI primitives
.out	Output of CARRY, CASCADE, EXP, GLOBGAL, LCELL, OPNDRN, SOFT, and TRI primitives

5 DESIGN EXAMPLES

This chapter describes some design examples illustrating the use of various design tools. Also it emphasizes textual design entry and a hierarchical approach to digital systems design. The first design is an electronic lock that is used to enter the key which consists of five decimal digits and unlock if the right combination is entered. The second design is a temperature control system that enables temperature control within a specified range in a small chamber by using a fan for cooling and a lamp for heating.

5.1 Electronic Lock

An electronic lock is a circuit that recognizes a 5-digit input sequence and indicates that the sequence is recognized using the unlock signal. A sequence of five 4-bit digits is entered using a hexadecimal keypad. If any of the digits in the sequence is incorrect, the lock resets and indicates that the new sequence should be entered from the beginning. This indication appears at the end of the sequence entry to increase the number of possible combinations that can be entered, making the task more difficult for a potential intruder.

The electronic lock consists of three major parts, as illustrated in the block diagram of Figure 5.1.

The input is a hexadecimal keypad and keypad encoder that accepts a keypress and produces the corresponding 4-bit binary code. The input sequence recognizer recognizes the sequence of five digits entered and produces an unlock signal if the sequence is correct. Output, which consists of three LEDs indicates that the lock is ready to accept a new sequence, or is accepting (active) a sequence of five digits, or the correct sequence has been entered (unlock), and a piezo buzzer output is activated whenever a key is correctly pressed. When a key is correctly pressed, the buzzer is held high for a short time interval of about 100 ms, corresponding to 5000 clock cycles of a 50kHz clock input.

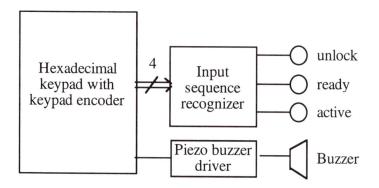

Figure 5.1 The Electronic Lock

The lock is supplied by an internal clock generator as shown in Figure 5.2. Its frequency is determined by the proper selection of an external resistor R and capacitors C1 and C2.

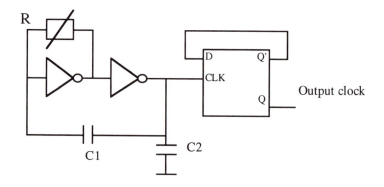

Figure 5.2 Internal clock implementation

5.1.1 Keypad encoder

The keypad encoder controls the hexadecimal keypad to obtain the binary code of the key being pressed. Different hexadecimal keypad layouts are possible, with the one represented in Figure 5.3 being the most frequently used.

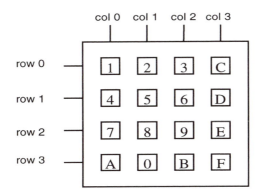

Figure 5.3 Hexadecimal keypad

The keypad encoder scans each row and senses each column. If a row and column are electrically connected, it produces the binary code for that key along with a strobe signal to indicate that the binary value is valid. The key press and produced binary code are considered valid if the debouncing circuitry, which is a part of the encoder, discovers the key has been held down for at least 126 clock cycles. When the key is settled, a strobe output is activated to indicate to the external circuitry (input sequence recognizer and buzzer driver) that the key press is valid. The debounce circuitry also ensures that the key does not auto-repeat if it is held down for a longer time.

The keypad encoder is shown in Figure 5.4. The purpose of this diagram is to visualize what will be described in an AHDL text file. The 4-bit counter is used to scan rows (two most significant bits) and select columns to be sensed (two low significant bits). When a key is pressed the counting process stops. The output of the 4-input multiplexer is used both as a counter enable signal as well as the signal indicating that a key is pressed. The value at the counter outputs represents a binary code of the key pressed. Different mappings of this code are possible if needed. The debouncing circuit gets the information that a key is pressed, and checks if it is pressed long enough. Its strobe output is used to activate the buzzer driver and indicates the input sequence recognizer that a binary code is present and the keypad encoder output is valid.

Debouncing circuitry is described by Example 5.1.

Example 5.1 Key Debouncing Circuitry

```
TITLE "Key debounce circuitry";

SUBDESIGN debounce
(
        clk             :INPUT;
        key_pressed     :INPUT; % Key currently pressed %
        strobe          :OUTPUT; % Key pressed for over 126 cycles
%
)

VARIABLE
        count[6..0]     :DFF; % 7-bit counter for key_pressed
cycles%

BEGIN
        count[].clk=clk;
        count[].clrn=key_pressed; % reset counter when no key
pressed %

        IF (count[].q<=126) & key_pressed THEN
                count[].d=count[].q+1;
        END IF;

        IF count[].q==126 THEN
                strobe = Vcc; % strobe produced only on 126 %
        ELSE
                strobe = GND;
        END IF;
END;
```

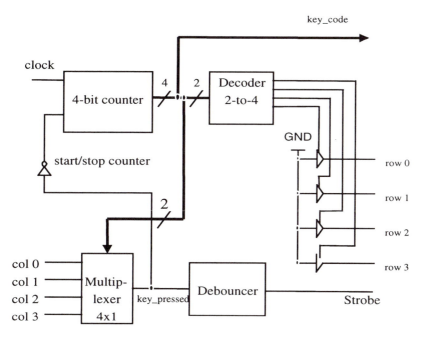

Figure 5.4 Hexadecimal keypad encoder

The remaining circuits used in the design of the keypad encoder, 4-to-1 multiplexer, 2-to-4 decoder, and the 4-bit counter have already been described in Chapter 4. Their designs are included in the keypad encoder design as shown in Example 5.2.

Example 5.2 Keypad Encoder.

```
TITLE "Keypad encoder";

INCLUDE   "4mux"; % Prototype for 4-to-1 multiplexer %
INCLUDE   "24dec"; % Prototype for 2-to-4 decoder %
INCLUDE   "4count"; % Prorotype for 4-bit counter %
FUNCTION debounce(clk, key_pressed)
RETURNS (strobe); % Prototype for debounce circuitry %
SUBDESIGN keyencode
(
        clk             : INPUT;
```

```
        col[3..0]        :INPUT; % Signals from keypad columns %
        row[3..0]        :OUTPUT; % Signals to keypad rows %
        key_code[3..0]   :OUTPUT; % Code of a key pressed %
        strobe           :OUTPUT; % Valid pressed key code %
)

VARIABLE
        key_pressed      :NODE; % Vcc when a key pressed %
        d[3..0]          :NODE; % Standard code for key %
        mux              :4mux; % Instance of 4mux %
        decoder          :24dec; % Instance of 24dec %
        counter          :4count; % Instance of 4count %
        opencol[3..0]    :TRI;   % Tristated row outputs %
BEGIN
        row[] = opencol[].out;
        opencol[].in = GND; % Inputs connected to GND %
        opencol[].oe = decoder.out[]; % Decoder drives keypad
                                                            rows %
        decoder.in[] = counter.(q3, q2);
        mux.d[] = col[];
        mux.sel[] = counter.(q1,q0);
        key_pressed = !mux.out;

% When a key is pressed its code appears on internal d[] lines%

        counter.clk = clk;
        counter.ena = !key_pressed;
        d[] = counter.q[];

% Code conversion for different keypad topologies %

        TABLE
                d[]      =>      key_code[]; % Any code mapping %
                H"0"     =>      H"1";
                H"1"     =>      H"2";
                 .
                 .
                 .
                H"F"     =>      H"F";
        END TABLE;

        strobe = debounce(clk, keypressed);
END;
```

5.1.2 Input Sequence Recognizer

An Input Sequence Recognizer (ISR) is a sequential circuit that recognizes a predefined sequence of five hexadecimal (4-bit) digits. The presence of a digit at the input of the ISR is signified by a separate strobe input. When the five digit sequence has been correctly received, the unlock output line is activated until the circuit is reset. When the first incorrect digit is received, the ISR is reset to its initial state. Assuming the sequence of digits to be recognized is $D_1D_2D_3D_4D_5$, the state machine used to describe the ISR is the state transition diagram in Figure 5.5. It has six states, where S0 is the initial state from which the ISR starts recognition of input digits.

After recognition of the required sequence, the circuit is returned to its initial state by the reset signal. The AHDL description of the ISR is shown in Example 5.3. Constants are used to define input digits, and can be easily changed as needed. In the case shown, the input sequence is 98765.

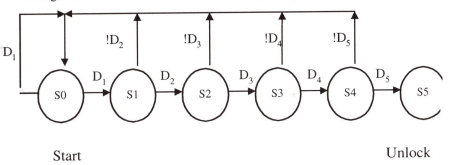

Figure 5.5 Input Sequence Recognizer State Transition Diagram

Example 5.3 Input Sequence Recognizer

```
TITLE "Input Sequence Recognizer";

CONSTANT D1 = H"9"; % First digit in sequence %
CONSTANT D2 = H"8"; % Second digit in sequence %
CONSTANT D3 = H"7"; % Third digit in sequence %
CONSTANT D4 = H"6"; % Fourth digit in sequence %
CONSTANT D5 = H"5"; % Fifth digit in sequence %
```

```
SUBDESIGN isr
(
      clk              :INPUT; % Indicate that the input code is
valid %
      i[3..0] :INPUT; % 4-bit binary code of input digit %
      rst              :INPUT; % Reset signal %
      unlock :OUTPUT; % Indicate correct input sequence %
)

VARIABLE
      sm        :MACHINE WITH STATES (s0, s1, s2, s3, s4, s5);

BEGIN

         sm.(clk, reset) = (clk, rst);
         CASE sm IS
         WHEN s0 =>
                 unlock = GND;
                 IF i[]==D1 THEN
                         sm = s1;
                 ELSE
                         sm = s0;
                 END IF;
         WHEN s1 =>
                 unlock = GND;
                 IF i[]==D2 THEN
                         sm = s2;
                 ELSE
                         sm = s0;
                 END IF;
         WHEN s2 =>
                 unlock = GND;
                 IF i[]==D3 THEN
                         sm = s3;
                 ELSE
                         sm = s0;
                 END IF;
         WHEN s3 =>
                 unlock = GND;
                 IF i[]==D4 THEN
                         sm = s4;
                 ELSE
                         sm = s0;
```

```
                 END IF;
     WHEN s4 =>
             unlock = GND;
             IF i[]==D5 THEN
                     sm = s5;
             ELSE
                     sm = s0;
             END IF;
     WHEN s5 =>
             unlock = Vcc;
             sm = s5;
     END CASE;
END;
```

This ISR design can be enhanced by adding states and outputs that make it safer and more user friendly. Additional outputs can be used to inform the user that the ISR is in its initial state, and that more digits must be entered. However, if we add the information that the ISR has returned to the initial state, the user easily knows which digit is incorrect and it reduces the number of possible combinations drastically. In order to overcome this, the number of possible combinations can be increased by increasing the number of states and always accepting five digits before resetting without informing the user which digit in the sequence was incorrect. One possible solution is illustrated by the state transition diagram in Figure 5.6. The states F1, F2, F3, and F4 are introduced to implement described behavior. By "x" we denote any input digit.

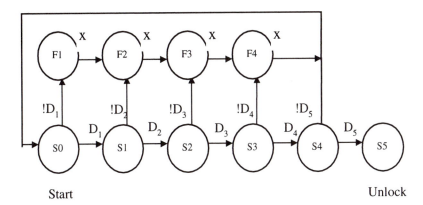

Figure 5.6 Modified ISR state transition diagram

The modified AHDL design of the ISR is given by Example 5.4.

Example 5.4 Modified Input Sequence Recognizer

```
TITLE "Modified Input Sequence Recognizer";

CONSTANT D1 = H"9"; % First digit in sequence %
CONSTANT D2 = H"8"; % Second digit in sequence %
CONSTANT D3 = H"7"; % Third digit in sequence %
CONSTANT D4 = H"6"; % Fourth digit in sequence %
CONSTANT D5 = H"5"; % Fifth digit in sequence %


SUBDESIGN modisr
(
        clk             :INPUT; % Indicate that the input code is
valid %
        i[3..0]         :INPUT; % 4-bit binary code of input digit %
        rst             :INPUT; % Reset signal %
        start           :OUTPUT; % New sequence to be entered %
        more            :OUTPUT; % More digits to be entered %
        unlock :OUTPUT; % Indicate correct input sequence %
)


VARIABLE
        sm              :MACHINE WITH STATES (
                             s0,s1,s2,s3,s4,s5,f1,f2,f3,f4);

BEGIN
        sm.(clk, reset) = (clk, rst);

        CASE sm IS

        WHEN s0 =>
                start= Vcc; more = GND; unlock = GND;
                IF i[]==D1 THEN
                        sm = s1;
                ELSE
                        sm = f1;
```

```
                END IF;
      WHEN s1 =>
              start= GND; more = Vcc; unlock = GND;
              IF i[]==D2 THEN
                      sm = s2;
              ELSE
                      sm = f2;
              END IF;
      WHEN s2 =>
              start= GND; more = Vcc; unlock = GND;
              IF i[]==D3 THEN
                      sm = s3;
              ELSE
                      sm = f3;
              END IF;
      WHEN s3 =>
              start= GND; more = Vcc; unlock = GND;
              IF i[]==D4 THEN
                      sm = s4;
              ELSE
                      sm = f4;
              END IF;
      WHEN s4 =>
              start= GND; more = Vcc; unlock = GND;
              IF i[]==D5 THEN
                      sm = s5;
              ELSE
                      sm = f5;
              END IF;
      WHEN s5 =>
              start= GND; more = GND; unlock = Vcc;
              sm = s5;
      WHEN f1 =>
              start= GND; more = Vcc; unlock = GND;
              sm = f2;
      WHEN f2 =>
              start= GND; more = Vcc; unlock = GND;
              sm = f3;
      WHEN f3 =>
              start= GND; more = Vcc; unlock = GND;
              sm = f4;
      WHEN f4 =>
              start= GND; more = Vcc; unlock = GND;
```

```
        sm = s0;
    END CASE;
END;
```

5.1.3 Piezo Buzzer Driver

When a key sequence is correctly pressed, the buzzer is held high for approximately 100ms, which is 5000 clock cycles of a 50kHz clock input. The beep signal is used to drive the buzzer and indicate the user that the key sequence has been entered correctly. The AHDL description of the piezo buzzer driver is shown in Example 5.5.

Example 5.5 Piezo Buzzer Driver.

```
TITLE "Piezo Buzzer Driver";

SUBDESIGN beeper
(
        clk            : INPUT;
        strobe : INPUT; % Valid key press %
        beep            : OUTPUT; % Piezo buzzer output %
)

VARIABLE
        buzzer : SRFF; % Buzzer SR flip-flop %
        count[12..0] : DFF; % 13 bits for internal counter %

BEGIN
        count[].clk = clk;
        buzzer.clk = clk;
        buzzer.s = strobe; % set FF when key pressed %
        count[].clrn = buzzer.q; % clear counter when buzzer
stops %

        IF buzzer.q AND (count[].q < 5000) THEN
                % increment counter %
                count[].d = count[].q + 1;
        END IF;
```

```
        IF (count[].q == 5000) THEN
                % reset buzzer output %
                buzzer.r = Vcc;
        END IF;

        beep = buzzer.q;
END;
```

5.1.4 Integrated Electronic Lock

Once all components are designed, we can integrate them into the overall design of the electronic lock. The integrated lock is flexible in respect to different required input sequences (passwords) and different keypad topologies. The final design is represented in Example 5.6.

Example 5.6 Electronic Lock.

```
TITLE "Electronic Lock";

% Prototypes of components %
FUNCTION keyencode(clk, col[3..0])
        RETURNS (row[3..0], key_code[3..0], strobe);

FUNCTION misr(clk, i[3..0], rst)
        RETURNS (start, more, unlock);

FUNCTION beeper(clk, strobe)
        RETURNS (beep);

SUBDESIGN lock
(
        clk             : INPUT;
        reset           : INPUT;
        col[3..0]       : INPUT;
        row[3..0]       : OUTPUT;
        start           : OUTPUT;
        more            : OUTPUT;
        unlock          : OUTPUT;
        buzzer          : OUTPUT;
```

```
)

VARIABLE
        key_code[3..0] :NODE;
        strobe          :NODE;

BEGIN
        buzzer = beeper(clk, strobe);
        (row[], key_code[], strobe) = keyencode(clk, col[]);
        (start, more, unlock) = misr(strobe, key_code[], reset);

END;
```

5.2 Temperature Control System

The temperature control system of this example is capable of keeping the temperature inside a small chamber within a required range between 20^0 and 99.9^0C. The required temperature range can be set with the user interface, in our case the hexadecimal keypad described in section 5.1. The temperature of the chamber is measured using a temperature sensor. When the temperature is below the lower limit of the desired range, the chamber is heated using an AC lamp. If it is above the upper limit of the range, the chamber is cooled using a DC fan. When the temperature is within the range, no control action is taken. The temperature is continuously displayed on a 3-digit hexadecimal display to one decimal place (for instance, 67.4^0C). Additional LEDs are used to indicate the state of the control system. The overall approach to the design uses decomposition of the required functionality into several subunits and then their integration into the control system. A simplified diagram of the temperature control system is shown in Figure 5.7. It is divided into the following subunits:

- Temperature sensing circuitry provides the current temperature in digital form.

- Keypad circuitry is used for setting high and low temperature limits and for operator-controller communication.

- Display driving circuitry is used to drive 3-digit 7-segment display.

- DC fan control circuitry is used to switch on and off the DC fan.

- AC lamp control circuitry is used to switch on and off AC lamp.

- Control unit circuitry implements the control algorithm and provides synchronization of operations carried out in the controller. Since the purpose of the design is not to show an advanced control algorithm, a simple on/off control will be implemented.

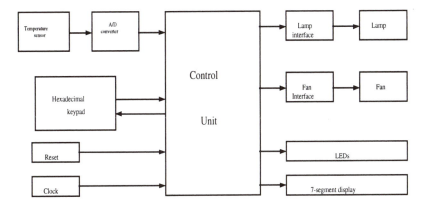

Figure 5.7 Block diagram of the temperature control system

The goal is to implement all interface circuits between the analog and digital parts of circuit, including the control unit, in an FPLD. Our design will be described by graphic and text entry tools provided in Max+PLUS II design environment.

5.2.1 Temperature Sensing and Measurement Circuitry

A sensor that measures the temperature inside the chamber is a transducer which converts a physical quantity, the temperature, to another more manageable quantity, a voltage. In our case we decided to use a common LM35 temperature sensor since it is a precision integrated circuit whose output voltage is linearly proportional to the temperature in Celsius degrees. Its important features are a large temperature range and satisfactory accuracy. Linear $+10\text{mV}/^0\text{C}$ scale factor in our case provides the output voltage between 0.2V and 0.99V for the temperature range from 20^0C and 99.9^0C. The analog voltage at the output of the sensor need to be converted to a digital form by an analog-to-digital (A/D) converter to be processed by the control unit.

A/D converters are available from many manufacturers with a wide range of operating characteristics. For our application, we have chosen the ADC0804 CMOS A/D converter which uses the successive approximation conversion method. Its pinout is illustrated in Figure 5.8.

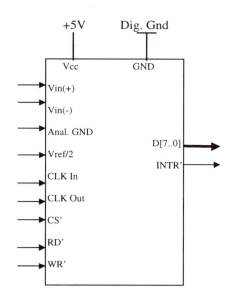

Figure 5.8   ADC0804

The ADC0804 major features are:

- Two analog inputs allowing differential inputs. In our case only one, Vin (+), is used. The converter uses Vcc=+5V as its reference voltage.

- Analog input voltage is converted to an 8-bit digital output which is tri-state buffered. The resolution of the converter is 5V/255=19.6mV.

- In our case Vref/2 is used as an input to reduce the internal reference voltage and consequently the analog input range that converter can handle. In this application, the input range is from 0 to 1V, and Vref/2 is set to 0.5V in order to achieve better resolution which is 1V/255=3.92mV

- The externally connected resistor and capacitor determine the frequency of an internal clock which is used for conversion (in our case the frequency of cca 600kHz provides the conversion time of 100us).

- The chip Select, CS', signal must be in its active-low state for RD' or WR' inputs to have any effect. With CS' high, the digital outputs are in the high-impedance state, and no conversion can take place. In our case this signal is permanently active.

- RD' (Read, Output Enable) is used to enable the digital output buffers to provide the result of the last A/D conversion. In our case this signal is permanently active.

- WR' (Write, Start of Conversion) is activated to start a new conversion.

- INTR' (End of Conversion) goes high at the start of conversion and returns low to signal the end of conversion.

It should be noticed that the output of the A/D converter is an 8-bit unsigned binary number which can be used by the control unit for further processing. However, this representation of the current temperature is not suitable for display purposes, in which case we prefer a BCD-coded number, or for comparison with the low and high temperature limits, which are entered using the keypad and are also in the BCD-coded form. This is the reason to convert the temperature into a BCD-coded format and do all comparisons in that format. This may not be the most effective way of manipulating temperatures. Therefore other possibilities can be explored.

The control unit is responsible for starting and coordinating the activities of A/D conversion, sensing the end of the conversion signal, and generating the control signals to read the results of the conversion.

5.2.2 Keypad Control Circuitry

The high and low temperature limits are set using a hexadecimal keypad as shown in Section 5.1. Besides digits 0 trough 9 used to enter values of temperature, this keypad provides the keys, A through F, which can be used as function keys. In our case we assign the functions to functional keys as shown in Table 5.1.

Table 5.1 Key Functions.

Key Code	Function
A	SL - Set low temperature limit
B	SH - Set high temperature limit
C	DL -Display low temperature limit
D	DH - Display high temperature limit
E	Enter; Return to running mode

Each time, a function key to set the temperature is pressed first, and then the value of the temperature limit. This pressing brings the system into the setting mode of operation. The entry is terminated by pressing E and the system is returned to the running mode of operation. When an incorrect value of the temperature limit is entered, the reset button is to be pressed to bring the system into its initial state and allow repeated entry of the temperature limits. In the case of pressing function keys for displaying temperature limits, the system remains in the running mode, but a low or high limit is displayed. The low and high temperature limits have to be permanently stored in internal control unit registers and available for comparison to the current measured temperature. In order to change the low and high limit, new values have to be passed to the correct registers. The overall structure of the keypad control circuitry is illustrated in Figure 5.9. In addition to the already shown keypad encoder, this circuitry contains a functional key decoder which recognizes the pressing of any functional key and activates the corresponding signal used by the control unit, and two 12-bit registers used to store two 3-digit BCD-coded values of the high and low temperature limit.

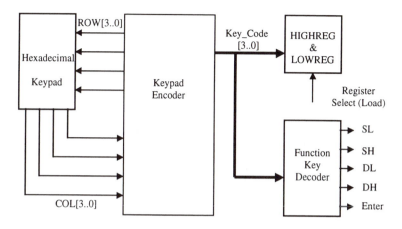

Figure 5.9 Keypad Control Circuitry

The values are entered into these registers digit-by-digit in a sort of FIFO arrangement, and are available on register output in the parallel form, as illustrated in Figure 5.10. As soon as the A or B key is pressed, the corresponding register is cleared, zeros displayed on the 7-segment displays, and the process of entering a new value can start. The register select signal is used to select to which of two registers, HIGH or LOW, will be forwarded.

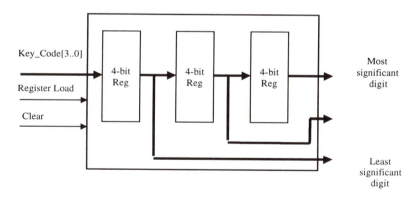

Figure 5.10 Low and High Temperature Limit Registers

5.2.3 Display Circuitry

Display circuitry provides the display for any one of the possible temperatures which are in the range from 00.0°C to 99.9°C. We decided to use 7-segment displays to display individual digits, as shown in figure 5.11.

Figure 5.11 Temperature Display.

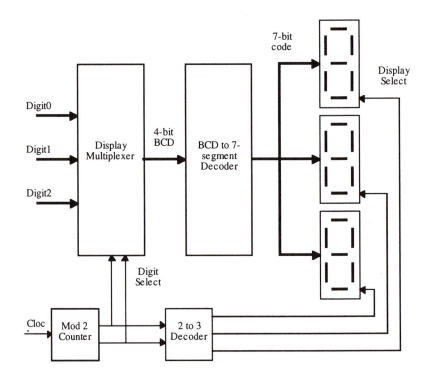

Figure 5.12 Display circuitry

Since the 7-segment display accepts on its inputs a 7-segment code, it is required to convert the BCD coded digits to 7-segment coded digits. The simplest method is to use a decoder for each digit and operate each independently. The other possibility is to use time division multiplexing. The displays are selected at the same time as the digits are to be passed through the multiplexer, as illustrated in Figure 5.11. The input to the display circuitry is 12 bits, representing the 3-digit BCD-coded temperature. A modulo 2 counter selects a digit to display and at the same time a 2-to-3 decoder selects a 7-segment display.

Selection of the currently displayed temperature is done by a 3-to-1 multiplexer illustrated in Figure 5.12. Inputs to this multiplexer are the current temperature and the low and high temperature limits. The control unit selects, upon request, the temperature which will appear on the 7-segment displays.

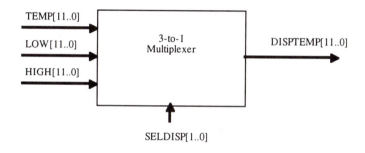

SELDISP[1..0]

Figure 5.12 Multiplexer for selection of the temperature to display

5.2.4 Fan and Lamp Control Circuitry

A control circuit using a highside driver to control the switching of the DC motor of a fan is used. An example of such a driver is National LM1051. Its worst case switching times for both turn on and off are 2us. A brushless DC fan with a 12V nominal operating voltage and low power consumption was chosen. The DC fan control circuitry is controlled by only one signal from the control unit providing on and off switching of the fan, as it is illustrated in Figure 5.13.

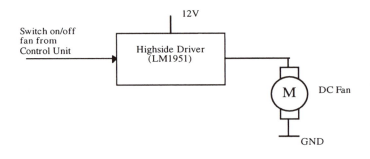

Figure 5.13 DC fan control circuitry

We choose a 150W infra heat AC lamp with instantaneous response and no warm up or cool down delay for heating up the chamber. To control the lamp with a digital signal, we used a triac and a zero crossing optically-coupled triac driver. The triac is a 3-terminal AC semiconductor switch which is triggered into conduction when a low energy signal is applied to its gate terminal. An effective method of controlling the average power to a load through the triac is by phase control. That is, to apply the AC supply to the load (lamp) for a controlled fraction of each cycle. In order to reduce noise and electromagnetic interference generated by the triac, we used a zero crossing switch. This switch ensures that AC power is applied to the load either in full or half cycles. The triac is gated at the instant the sine wave voltage is crossing zero. An example of a zero crossing circuit is the optoisolator MOC3031, which is used in interfacing with AC powered equipment. The entire AC lamp control circuitry is illustrated in Figure 5.14. It is controlled by signal generated from the control unit to switch the lamp on or off.

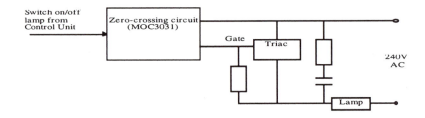

Figure 5.14 AC Lamp Control Circuitry

5.2.5 Control Unit

The control unit is the central point of the design. It provides proper operation of the temperature control system in all its modes including temperature sensing, switching between modes, communication with the operator, control of data flow, and data processing in the data path of the circuit. The main inputs to the control unit are current temperature (including synchronization signals for temperature measurement), high and low temperature limit, and the inputs from the keypad. The main outputs from the control unit are signals to select different data paths (to compare current temperature with the low and high temperature limit), signals to control on/off switching of the fan and lamp, signals to control temperature measurement, and signals to control the displays (7-segment and LEDs). The operation of the control unit is illustrated by the flow diagram in Figure 5.15.

After power-up or reset, the control unit passes the initialization phase (where all registers are cleared), low temperature limit, LOW, is selected for comparison first, and current temperature, TEMP, is selected to be displayed. After that, the control unit checks the strobe signal of the keypad. If there is no strobe signal, the control unit enters the running mode of operation which includes a new A/D conversion cycle to get the value of the TEMP, comparison of the TEMP with the LOW in order to turn on lamp, or comparison of the TEMP with the high temperature limit, HIGH, in order to turn on fan. This process is repeated until a key press is detected. When a key press is detected, the control unit recognizes the code of the key and appropriate operations are carried out. Upon completion of these operations, the system is returned to the running mode.

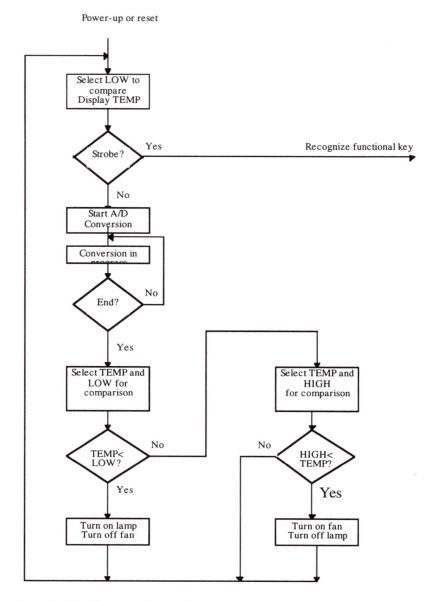

Figure 5.15 (a) Control unit flow diagram

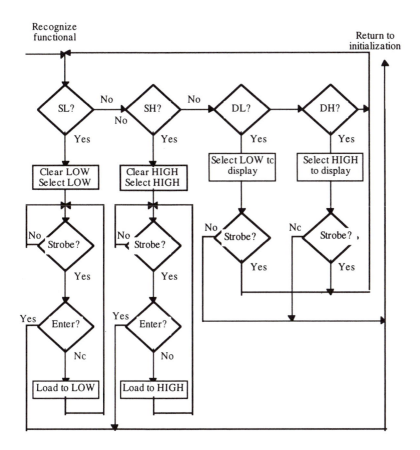

Figure 5.15 (b) Control Unit Flow Diagram

The following state machine states implement the control unit behavior:

S0 - Initialization, upon power-up or reset

S1 - Start A/D conversion

S2 - A/D conversion in progress

S3 - Comparison of current temperature and low temperature limit

S4 - Comparison of current temperature and high temperature limit

S5 - Setting low temperature limit

S6 - Setting high temperature limit

S7 - Display low temperature limit

S8 - Display high temperature limit

Finally, we identify inputs and outputs of the control unit. If we assume all data transfers and processing are done within the data path, then the inputs and outputs of the control unit are identified as shown in Table 5.2.

Table 5.2 Control Unit Inputs and Outputs.

Input	Function
END	End of A/D conversion - active LOW signal
STROBE	HIGH pulse from Keypad encoder indicating valid key press
SL, SH, DL, DH, Enter	Indication of active functional keys received from Keypad encoder
AGEB	Indication that temperature A is greater or equal to temperature B
ALTB	Indication that temperature A is lower than temperature B
Output	Function
START	LOW pulse sent to the WR' pin of A/D converter to enable start of the next conversion
SET_LOW	Signal indicating that the control system is in setting the LOW temperature limit - used to switch on LOW LED
SET_HIGH	Signal indicating that the control system is in setting the HIGH temperature limit - used to switch on HIGH LED
CLR_LOW	Signal for clearing the value of the LOW register
CLR_HIGH	Signal for clearing the value of the HIGH register

LD_LOW	Signal used to load the LOW register with the next BCD-coded digit received from keypad at the presence of the STROBE signal
LD_HIGH	Signal used to load the HIGH register with the next BCD-coded digit from the keypad at the presence of the STROBE signal
SELHILO	Select signal of the multiplexer used to select either HIGH or LOW temperature limit to be forwarded to the comparator and compared to the current temperature TEMP
SELDISP[1..0]	Select signal of the multiplexer used to select which signal will be displayed on the 7-segment displays
FAN_ON	Signal used to turn on the DC fan and at the same time to turn off the AC lamp
LAMP_ON	Signal used to turn on the AC lamp and at the same time to turn off the DC fan

5.2.6 Temperature Control System Design

Our approach to the overall design of the temperature control system follows a traditional line of partitioning into separate design of the data path and the control unit, and their easy integration into the target system. This approach is illustrated in Figure 5.16.

The data path provides all facilities to exchange data with external devices, while the control unit uses internal input control signals generated by the data path, external input control signals to generate internal control outputs that control data path operation, and external control outputs to control external devices.

The data path of the temperature control system is made up of all circuits that provide: interconnections, the data path, and the data transformation.

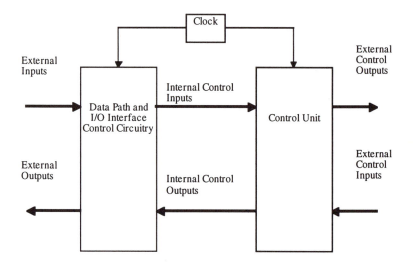

Figure 5.16 Global approach to temperature control system design

1. Interconnections between main input subunits (hexadecimal keypad and A/D converter) and main output subunits (7-segment displays, LED displays, and control signals that switch on and off DC fan and AC lamp)

2. Data paths to and from internal registers that store low and high temperature limits.

3. Data transformations and processing (binary to BCD-coded format of the current temperature, comparison of current temperature to low and high temperature limits)

   The temperature control system data path, together with the circuits that control input and output devices is shown in Figure 5.17. All control signals, except those for clearing the LOW and HIGH registers, generated by the control unit are also shown.

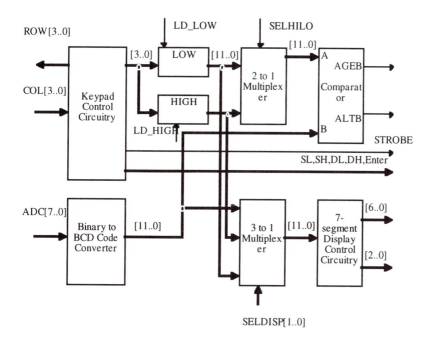

Figure 5.17 Temperature Control System Data Path with interfacing circuitry

The designs of almost all subunits of the data path have been already shown in various parts of the book. Remaining details and integration to the overall data path are left to the reader as an exercise.

Finally, we come to the design of the control unit explained in Section 5.2.5. This design is easily transformed into the corresponding AHDL description of the state machine. The AHDL design file of the control unit with appropriate comments is shown as Example 5.7.

Example 5.7 Temperature Control System Unit.

```
TITLE "Temperature Control System Control Unit";

SUBDESIGN ctrlunit
(
        clk                     : INPUT;
        reset                   : INPUT;
```

```
        sl,sh,dl,dh,enter      :INPUT; % functional key activation
inputs %
        endc                    :INPUT; % end of a/d conversion %
        strobe          :INPUT; % keypress valid signal %
        ageb            :INPUT; % compared temp >= current temp %
        altb            :INPUT; % compared temp < current temp %
        start                   :OUTPUT; % start a/d conversion %
        set_low,set_high        :OUTPUT; % activate external leds %
        clr_low,clr_high        :OUTPUT; % clear low, high register
%
        ld_low,ld_high :OUTPUT; % load data into register file %
        selhilo         :OUTPUT; % select reg to be compared %
        seldisp[1..0]   :OUTPUT; % select temperature to display %
        fan_on,lamp_on :OUTPUT; % turn on/off fan/lamp %
        oe_adc          :OUTPUT; % read data from a/dc %
)


VARIABLE
        sm: MACHINE WITH STATES
                (s0,s1,s2,s3,s4,s5,s6,s7,s8);


BEGIN
        sm.(clk, reset)=(clk, reset);

        CASE sm IS

        WHEN s0 =>
                set_low = GND;
                set_high = GND;
                selhilo=GND; % select low to compare %
                seldisp[]=B"00"; % select temp to display %
                IF !strobe THEN
                        sm = s1;
                ELSIF strobe & sl THEN
                        clr_low = Vcc;
                        seldisp[] = B"01"; % low to display %
                        set_low = Vcc; % turn on set_low led %
                        sm = s5;
                ELSIF strobe & sh THEN
                        clr_high = Vcc;
                        seldisp[] = B"10"; % high to display %
                        set_high = Vcc; % turn on set_high led %
                        sm = s6;
```

```
        ELSIF strobe & dl THEN
                seldisp[] = B"01"; % low to display %
                sm = s7;
        ELSIF strobe & dh THEN
                seldisp[] = B"10"; % high to display %
                sm = s8;
        ELSE
                sm = s0; % functional key not pressed %
        END IF;

WHEN s1 =>
        start = Vcc; % start a/dc %
        sm = s2;

WHEN s2 =>
        IF endc THEN
                oe_adc = Vcc; % read a/dc %
                sm = s3;
        ELSE
                sm = s2;
        END IF;

WHEN s3 =>
        IF ageb THEN % low >= temp %
                lamp_on = Vcc;
                sm = s0;
        ELSE
                sm = s4;
        END IF;

WHEN s4 =>
        selhilo = Vcc; % select high to compare %
        IF altb THEN % high < temp %
                fan_on = Vcc;
                sm = s0;
        ELSE
                sm = s0;
        END IF;

WHEN s5 =>
        IF !strobe THEN
                sm = s5; % wait for key press %
        ELSIF strobe & enter THEN
```

```
                           sm = s0; % enter pressed - new value
entered %
                ELSIF strobe & !enter THEN
                        ld_low = Vcc; % new digit of low %
                        sm = s5;
                END IF;

        WHEN s6 =>
                IF !strobe THEN
                        sm = s6; % wait for key press %
                ELSIF strobe & enter THEN
                        sm = s0; % enter pressed - new value
entered %
                ELSIF strobe & !enter THEN
                        ld_high = Vcc; % new digit of high %
                        sm = s6;
                END IF;

        WHEN s7 =>
                seldisp[] = B"01"; % low to display %
                IF strobe & dl THEN
                        sm = s7; % dl kept pressed %
                ELSE
                        sm = s0; % dl released %
                END IF;

        WHEN s8 =>
                seldisp[] = B"10"; % high to display %
                IF strobe & dh THEN
                        sm = s8; % dh kept pressed %
                ELSE
                        sm = s0; % dh released %
                END IF;
        END CASE;
END;
```

We have done compilation experiments with different target devices to investigate resource utilization. In the first experiment the Max+PLUS II compiler automatically selected the FLEX 8282 device. The design of the control unit required 33 out of 282 logic cells occupying 15% of this basic resource of the lowest capacity FLEX 8000 device. In the second experiment, the compiler automatically selected a MAX 7032 device. The design required 17 out of 32 macrocells and 13 shareable expanders occupying 53% of the lowest capacity MAX 7000 device.

6 SIMP - A SIMPLE CUSTOM-CONFIGURABLE MICROPROCESSOR

In this chapter we will discuss the design of a simple 16-bit custom configurable microprocessor called SimP. The SimP can be considered the core for various user specific computing machines. It consists of a set of basic microprocessor features that can be used without any changes for some simple applications, or can be extended by the user in many application specific directions. Extensions can be achieved by adding new instructions or other features to the SimP's core, or by attaching functional blocks to the core without actually changing the core.

The design of SimP is almost completely specified in AHDL, and the methodology of adding new features is shown at the appropriate places.

6.1 Basic Features

The basic features of the SimP core are:

- 16-bit external data bus and 12-bit address bus that enable direct access to up to 4096 16-bit memory locations

- Two programmer visible 16-bit working registers, called A and B registers, which are used to store operands and results of data transformations

- memory-mapped input/output for communication with the input and output devices

- it is basically a load/store microprocessor with a simple instruction cycle consisting of four machine cycles per each instruction; all data transformations are performed in working registers

- support of direct and the most basic stack addressing mode

- definable custom instructions and functional blocks which execute custom instructions can be added

- it is implemented in a CPLD from the FLEX 8000 family, with an active serial configuration scheme

- physical pin assignments can be changed to suit the PCB layout.

6.1.1 Instruction Formats and Instruction Set

The SimP instructions have very simple formats. All instructions are 16-bits long and require one memory word. In the case of direct addressing mode, 12 lower instruction bits represent an address of the memory location. All other instructions for basic data processing, program flow control, and control of processor flags use implied addressing mode. The core instruction set is illustrated in Table 6.1.

Table 6.1 SimP Instruction Set

Instruction mnemonic	Function
LDA	A ← M[address]
LDB	B ← M[address]
STA	M[address] ← A
STB	M[address] ← B
JMP	PC ← address
JSR	stack ← PC, PC ← address, SP ← SP-1
ADD	A ← A + B
AND	A ← A and B
CLA	A ← 0

PSHA	stack ← A, SP ← SP-1
PULA	SP ← SP+1, A ← stack
CLB	B ← 0
CMB	B ← B'
INCB	B ← B + 1
DECB	B ← B - 1
CLflag	flag ← 0 (flag can be Carry,Zero)
ION	IEN ← 1, Enable interrupts
IOF	IEN ← 0, Disable interrupts
SZ	If Z=1, PC ← PC + 1
SC	If C=1, PC ← PC + 1
RET	SP ← SP+1, PC ← stack

All memory reference instructions use either direct or stack addressing mode and have the format as shown below:

Opcode(4)	Address(12) or not used

The four most significant bits are used as the operation code (opcode) field. The opcode field is 4-bits long and can specify up to 16 different instructions. Twelve least significant bits are used as an address for instructions with direct addressing mode or they have no meaning for instructions using stack (implicit) addressing mode. Although the stack pointer (SP) is present, it is not directly visible to programmer. It is initialized at the power-up of the microprocessor to the value FF0 (hex) and subsequently changes its value as instructions using stack or external interrupts occur.

Memory reference instructions with the direct and stack addressing modes are assigned the opcodes as shown in Table 6.2.

Table 6.2 Opcode Instruction Mnemonics

Opcode i[15..12]	Instruction Mnemonic
0000	LDA
0001	LDB
0010	STA
0011	STB
0100	JMP
1000	JSR
1010	PSHA
1100	PULA
1110	RET

Instructions in direct addressing mode, (that do not use the stack) have the most significant bit i[15] equal to 0. Those that use the stack have bit i[15] equal to 1. Although it is not shown here, the instructions which belong to the register reference instructions and are not using the stack have the bit i[15] equal to 0 and four most significant bits equal to 7 (hex), and instructions that operate on user-specific functional blocks have the bit i[15] equal to 1 and four most significant bits equal to F (hex). The remaining SimP core instructions have the following instruction formats

Opcode(8)	Not used

and are assigned the opcodes as shown in Table 6.3.

Table 6.3 Instruction Mnemonics

Opcode i[15..8]	Instruction Mnemonic
0111 0001	ADD
0111 0010	AND
0111 0011	CLA
0111 0100	CLB
0111 0101	COMB
0111 0110	INCB

0111 0111	DECB
0111 1000	CLC
0111 1001	CLZ
0111 1010	ION
0111 1011	IOF
0111 1100	SC
0111 1101	SZ

Register reference instructions operate on the contents of working registers (A and B), as well as on individual flag registers used to indicate different status information within the processor or to enable and disable interrupts.

Finally, program flow control instructions are used to change program flow depending on the results of current computation.

Besides the shown instructions, the SimP provides instructions that invoke different application specific functional blocks. These instructions are designated with instruction bits i[15] and i[12] set to 1. The individual instructions are coded by the least significant bits i[7..0].

6.1.2 Register Set

SimP contains a number of registers that are used in performing microoperations. Some are accessible to the user, and the others are used to support internal operations. All user visible registers are shown in Figure 6.1.

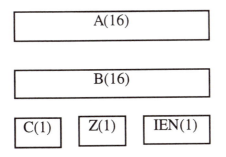

Figure 6.1 User Visible Registers

Registers that are not visible to the user include 12-bit program counter, PC, and a 12-bit stack pointer, SP. The program counter is used during instruction execution to point to the next instruction. The stack pointer is used to implement the subroutine and interrupt call, and return mechanism (to save and restore return addresses). It also supports the execution of "push to the stack" and "pull from the stack" instructions. At system power-up, the program counter is loaded with the value H"000" and the stack pointer with the value H"FF0". The stack grows towards the lower addresses. Two further registers, the program counter temporary register (TEMP), and the auxiliary stack pointer register (ST), are neither directly nor indirectly accessible by the user. They are used in internal operations to save values (copies) of the program counter and stack pointer and provide a very simple, uniform instruction execution cycle. As we shall see, all instructions execute in exactly four machine (clock) cycles, thus providing eventual performance improvement using pipelining in an advanced version of the microprocessor.

6.2 Processor Data Path

Overall SimP structure is presented in the simplified data path of Figure 6.2. The processor contains two internal buses: a 16-bit data bus and a 12-bit address bus. The data bus is connected to the external pins and enables easy connection with external memory up to 4,096 16-bit words or to the registers of input and output devices in a memory mapped scheme. The external data bus appears as bi-directional, while internally it provides separated input and output data bus lines. The address bus is available only for internal register transfers and enables two simultaneous register transfers to take place.

Additional registers not visible to the user appear in the in the internal structure of the data path. They are the 16-bit instruction register (IR) and the 12-bit address register (AR). The instruction register is connected to the instruction decoder and provides input signals to the control unit. The details of the use of all registers will be explained in upcoming sections.

The Arithmetic-Logic Unit (ALU) performs simple arithmetic and logic operations on 16-bit operands as specified in the instruction set. In its first version, the ALU performs only two operations, "addition" and "logical and." It can easily be extended to perform additional operations. Some data transformations, such as incrementation and initialization of working registers, are carried out by the surrounding logic of working registers A and B.

Access to the external memory and input output devices is provided through multiplexers that are used to form buses. An external address is generated on the address lines, A[11..0], as the result of selected lines of the memory multiplexer, (MEMMUX). Usually, the effected address is contained in address register AR, but in some cases it will be taken from another source, stack pointer (SP) or auxiliary stack pointer (ST).

Two other multiplexers (used to form the address and data bus) are not shown in Figure 6.2. However, it is obvious that several registers or memory can be the source of data on both buses. Two multiplexers, the address bus multiplexer (ABUSMUX) and data bus multiplexer (DBUSMUX) are used to enable access to address and data bus, respectively. The only register that can be the source of data for both these buses is the program counter (PC). If the contents of the program counter is transferred by way of data bus, only the 12 least significant data lines are used for the actual physical transfer.

External memory can be both the source and destination in data transfers. This is determined by the memory control lines that specify either the memory read (MR) or memory write (MW) operation. Memory location (that takes place in data transfer) is specified by the value of the output of MEMMUX which in turn specifies the effective address.

All register transfers are initiated and controlled by the control unit. It carries out the selection of the data source for each of internal bus, destination of data transfer, as well as operations local to individual resources. For example, the control unit activates the memory read or write control line, initializes an individual register, performs such operations as the incrementation or decrementation of the contents of a register, and selects the operation of the arithmetic-logic unit. All register transfers are synchronized by the system clock and take place at the next clock tick.

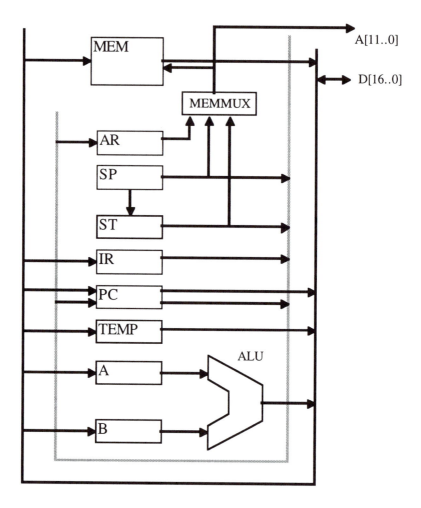

Figure 6.2 SimP Data Path

6.3 Instruction Execution

The SimP's core instructions are executed as sequences of microoperations. The basic instruction cycle contains all operations from the start to the end of an instruction. It is divided into three major steps that take place in the individual machine clock cycles denoted by T0, T1, T2, and T3.

1. Instruction fetch is when a new instruction is fetched from an external memory location pointed to by the program counter. It is performed in two machine cycles. The first cycle, T0, is used to transfer the address of the next instruction from the program counter to the address register. The second cycle T1 is used to actually read the instruction from the memory location into instruction register, IR.

2. Instruction decode is the recognition of the operation that has to be carried out and the preparation of effective memory address. This is done in the third machine cycle T2 of the instruction cycle.

3. Instruction execution is when the actual operation specified by the operation code is carried out. This is done in the fourth machine cycle T3 of instruction cycle.

Besides these three fundamental operations in each machine cycle, various auxiliary operations are also performed that enable each instruction to be executed in exactly four machine cycles. They also provide the consistency of contents of all processor registers at the beginning of each new instruction cycle.

Instructions are executed in the same sequence they are stored in memory, except for program flow change instructions. Besides this, the SimP provides a very basic single level interrupt facility that enables the change of the program flow based on the occurrence of external events represented by hardware interrupts. A hardware interrupt can occur at any moment since it is controlled by an external device. However, the SimP checks for the hardware interrupt at the end of each instruction execution and, in the case that the interrupt has been required, it sets an internal flip-flop called interrupt flip-flop (IFF). At the beginning of each instruction execution, SimP checks if IFF is set. If not set, the normal instruction execution takes place.

If the IFF is set, SimP enters an interrupt cycle in which the current contents of the program counter is saved on the stack and the execution is continued with the instruction specified by the contents of memory location called the interrupt vector (INTVEC).

The interrupt vector represents the address of the memory location which contains the first instruction of the Interrupt Service Routine (ISR), which then executes as any other program sequence. At the end of the ISR, the interrupted sequence, represented by the memory address saved on the stack at the moment of the interrupt acknowledgment, is returned to using the "ret" instruction.

The overall instruction execution and control flow of the control unit, including normal execution and interrupt cycle, is represented by the state flowchart of Figure 6.3. This flowchart is used as the basis of the state machine that defines the control unit.

Some other 1-bit registers (flip-flops) appear in the flowchart of Figure 6.3. First is the interrupt request flip-flop (IRQ). It is used to record the active transition on the interrupt request input line of the microprocessor. When the external device generates an interrupt request, the IRQ flip-flop will be set and, under the condition that interrupts are enabled, will cause the IFF flip-flop to be set. Consequently, the interrupt cycle will be initiated instead of normal instruction execution cycle. Control of the interrupt enable (IEN) flip-flop is carried out by programmer using instructions to enable or disable interrupts. Initially, all interrupts are enabled automatically. After recognition of an interrupt, further interrupts are disabled automatically. All other interrupt control is the responsibility of the programmer. During the interrupt cycle, the IRQ flip-flop will be cleared enabling new interrupt requests to be recorded. Also, interrupt acknowledgment information will be transferred to the interrupting device in the form of the pulse that lasts two clock cycles (IACK flip-flop is set in the machine cycle T1 and cleared in the cycle T3 of the interrupt cycle).

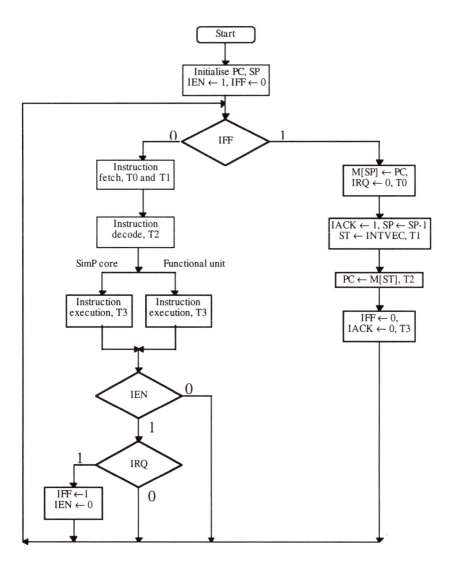

Figure 6.3 SimP's Control Flow Diagram

Now, we will describe the normal instruction execution cycle illustrated in the flowchart of Figure 6.4. In the first machine cycle, T0, the contents of the program counter is transferred to the address register. This register prepares the address of the memory location where the next program instruction is stored. The next machine cycle, T1, is first used to fetch and transfer an instruction to the instruction register to enable further decoding. In the same cycle, two other microoperations are performed.

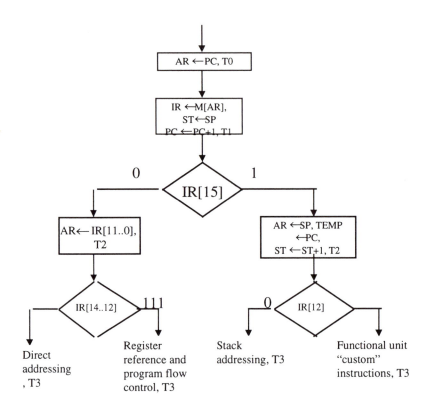

6.4 Instruction Execution Flowchart

The program counter is incremented to point to the next instruction which would be executed if there is no change in program flow. Also, the stack pointer

(SP) is copied into the ST. This is preparation for the possibility that the instruction uses stack addressing mode in the next machine cycle.

Register transfers that take place in the next machine cycle, T3, depend on the value of the most significant bit of the instruction fetched which is now bit IR[15]. If this value is equal to 0, direct or register addressing mode is used. If direct addressing mode is used, the lower 12 instruction bits, IR[11..0], represent the effective address which is used during the instruction execution step. Therefore, they are transferred to the address register preparing the effective address for the last machine cycle if needed. If IR[15] is equal to 1, two possibilities exist.

First, if the IR[12] bit is also 1, it is an instruction that executes a custom, application-specific instruction in a functional unit. Actions undertaken by the control unit for this case will be explained later. Otherwise, the instruction belongs to one using the stack addressing mode. To execute these instructions efficiently, preparation for all possible directions in which instruction execution can continue are done. First, the stack pointer is copied into the address register preparing for instructions that will push data onto the stack ("push" and "jump to subroutine" instruction) during the execution step. Second, the program counter is copied into the TEMP register to prepare for instructions that must save the contents of the program counter onto the stack and change the value of the program counter ("jump to subroutine" instruction). Finally, the ST register is incremented to prepare for instructions that pull data from the stack ("pull" and "ret" instructions). These steps also enable the proper updating (incrementing or decrementing) of the SP register in the T3 machine cycle, while the stack is accessed using the AR or ST register as the source of the effective address.

The instruction execution step performed in the T3 machine cycle for all instructions from the SimP's core is presented in the Table 6.4.

Table 6.4 Executed Step in the T3 Cycle.

Instruction	Microoperations (Register transfers)
LDA	A ← M[AR]
LDB	B ← M[AR]
STA	M[AR] ← A

STB	M[AR] ← B
JMP	PC ← IR[11..0]
JSR	M[AR] ← TEMP, PC ← IR[11..0], SP ← SP-1
ADD	A ← A+B
AND	A ← A and B
CLA	A ← 0
PSHA	M[AR] ← A, SP ← SP-1
PULA	A ← M[ST], SP ← SP+1
CLB	B ←0
CMB	B ← B'
INCB	B ← B+1
DECB	B ← B-1
CLflag	flag ← 0, flag = C,Z
ION	IEN ← 1
IOF	IEN ← 0
SZ	If Z=0 (PC ← PC+1); else nothing
SC	If C=0 (PC ← PC+1); else nothing
RET	PC ← M[ST], SP ← SP+1

6.4 SimP Implementation

Our approach to the SimP design follows a traditional path of digital system design partitioned into the data path and control unit parts as illustrated in Figure 6.5.

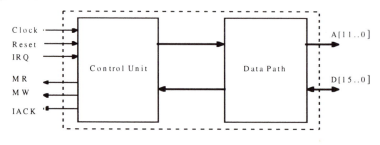

Figure 6.5 Basic Partition of SimP Design

The data path consist of all registers, interconnect structures (including various multiplexers), and data processing resources. The data path enables register transfers under the control of multiplexer selection signals and control signals of the registers, local operations on the contents of the registers, data transformations in the arithmetic-logic unit, and data exchange with the outside world (memory and input/output devices). From an external point of view it provides a 12-bit address bus and a 16-bit data bus. The control unit provides proper timing, sequencing and synchronization of microoperations, and activation of control signals at various points in the data path (as required by the microoperations). It also provides control signals which are used to control external devices such as memory operations and the interrupt acknowledgment signal. The operation of the control unit is based on information provided by the program (instructions fetched from memory), results of previous operations, as well as the signals received from the outside world. In our case the only signal received from the outside world is the interrupt request received from the interrupting device.

6.4.1 Data Path Implementation

In order to design all resources of the data path (Figure 6.1), we must first identify data inputs and data outputs of each resource, as well as operations that can be carried out and the control signals that initiate operations.

Program Counter

As an example, take the program counter (PC). Its data inputs, data outputs, and control signals are illustrated in Figure 6.6. By analyzing microoperations as well as resource usage, we see that the PC must provide 12-bit inputs from both internal address and data buses. These inputs are called PCA[11..0] and PCD[11..0] respectively. Consequently, the appropriate control signals which determine the input lines, called LDA and LDD, are provided as well. The PC must provide control signals that enable its initialization at system start-up (power-up), clear (CLR) and incrementation (INC).

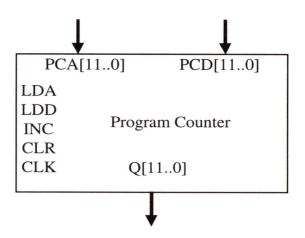

Figure 6.6 Program Counter

The AHDL design that describes the PC operation is given in Example 6.1.

Example 6.1 PC Operation.

```
TITLE "Program Counter pc";
SUBDESIGN pc
(
clk, clr,lda,ldd, inc               :INPUT;
pcd[11..0]                          :INPUT;
pca[11..0]                          :INPUT;
q[11..0]                            :OUTPUT;
)

VARIABLE
        ff[11..0]                   :DFF;

BEGIN
        ff[].clk=clk;
        q[]=ff[].q;
        ff[].clrn=!clr;

        IF ldd THEN
                ff[].d=pcd[];
        ELSIF lda THEN
                ff[].d=pca[];
        ELSIF inc THEN
```

```
                ff[].d=ff[].q+1;
        ELSE
                ff[].d=ff[].q;
        END IF;
END;
```

Stack Pointer

Another example of a register is the stack pointer (SP). It can be initialized to a specific value (FF0 [hex]) at the system start-up. As specified by its microoperations, the SP can be only initialized, incremented, and decremented. Its data inputs, outputs, and control lines are illustrated in Figure 6.7. The AHDL design describing the SP operation is given in Example 6.2.

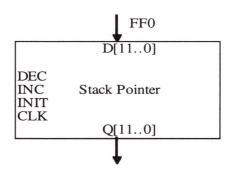

Figure 6.7 Stack Pointer

Example 6.2 SP Operation.

```
TITLE "Stack Pointer sp";

SUBDESIGN sp
(
clk, inc, dec, init        :INPUT;
q[11..0]                   :OUTPUT;
)

VARIABLE
```

```
        ff[11..0]                    :DFF;
        d[11..0]                     :NODE;

BEGIN
        ff[].clk=clk;
        q[]=ff[].q;
        d[]=H"FF0";

        IF init THEN
                ff[].d=d[];
        ELSIF dec THEN
                ff[].d=ff[].q-1;
        ELSIF inc THEN
                ff[].d=ff[].q+1;
        ELSE
                ff[].d=ff[].q;
        END IF;
END;
```

Working Registers

Working registers are used to store operands, results of operations, and carry out operations. The B register is slightly more complex and enables the microoperations of incrementing, decrementing, complementing of its contents, clearing, and loading contents from the input data lines. It is illustrated in Figure 6.8.

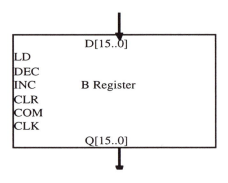

Figure 6.8 B Register

The operation of the B register is described in Example 6.3.

Example 6.3 Working Register b.

```
TITLE "Working Register b";

SUBDESIGN b
(
clk, clr,ld,inc,dec,com                     :INPUT;
d[15..0]                            :INPUT;
q[15..0]                           :OUTPUT;
)

VARIABLE
       ff[15..0]                          :DFF;

BEGIN
       ff[].clk=clk;
       q[]=ff[].q;
       ff[].clrn=!clr;

       IF ld THEN
                 ff[].d = d[];
       ELSIF inc THEN
               ff[].d=ff[].q+1;
       ELSIF dec THEN
               ff[].d=ff[].q-1;
       ELSIF com THEN
               ff[].d=!ff[].q;
       ELSE
               ff[].d=ff[].q;
       END IF;
END;
```

Other registers, including 1-bit registers that indicate the result of the most recent arithmetic-logic unit operation, are described by similar AHDL design files.

Arithmetic-Logic Unit

The Arithmetic Logic Unit (ALU) is designed in a hierarchical manner by first designing 1-bit ALU as a basic cell. The basic cell is then iterated 16 times in a structural model to produce the 16-bit ALU. The 1-bit ALU is described in Example 6.4.

Example 6.41-Bit ALU.

```
TITLE "1-bit alu alu1";
SUBDESIGN alu1
(
a,b,cin,als[1..0]                        :INPUT;
q,cout                    :OUTPUT;
)

BEGIN
        CASE als[1..0] IS
        WHEN B"00" =>
                q = a $ (b $ cin);
                cout=carry((a & b) # (a & cin) # (b & cin));
        WHEN B"01" =>
                q = a & b;
        WHEN B"10" =>
                q = a;
        WHEN B"11" =>
                q = b;
        END CASE;
END;
```

As its inputs the 1-bit ALU has two operand data inputs a and b and an input carry bit (cin) from the previous stage of multi-bit ALU, as well as two lines to select the operation, als[1..0]. The output results are present on the data output line (q) output carry (cout) and are used as input in the next stage of the multi-bit ALU. Operations performed by the 1-bit ALU are 1-bit addition, 1-bit "logical and", and transfer of input argument a or b. Transfer operations are needed because neither of the working registers has direct access to the data bus, but is accomplished through the ALU. The 16-bit ALU is designed using "pure" structural AHDL description as shown in Example 6.5.

Example 6.5 16-Bit ALU.

```
TITLE "16-bit alu alu16";

INCLUDE "alu1";

SUBDESIGN alu16
(
alusel[1..0], a[15..0], b[15..0], cin        :INPUT;
q[15..0], cout, zout                         :OUTPUT;
)
VARIABLE
        1alu[15..0]                          :ALU1;

BEGIN
        1alu[].a=a[];
        1alu[].b=b[];
        1alu[].als[]=alusel[];
        1alu[0].cin=cin;
        1alu[1].cin=soft(1alu[0].cout);
        1alu[2].cin=soft(1alu[1].cout);
        1alu[3].cin=soft(1alu[2].cout);
        1alu[4].cin=soft(1alu[3].cout);
        1alu[5].cin=soft(1alu[4].cout);
        1alu[6].cin=soft(1alu[5].cout);
        1alu[7].cin=soft(1alu[6].cout);
        1alu[8].cin=soft(1alu[7].cout);
        1alu[9].cin=soft(1alu[8].cout);
        1alu[10].cin=soft(1alu[9].cout);
        1alu[11].cin=soft(1alu[10].cout);
        1alu[12].cin=soft(1alu[11].cout);
        1alu[13].cin=soft(1alu[12].cout);
        1alu[14].cin=soft(1alu[13].cout);
        1alu[15].cin=soft(1alu[14].cout);
        cout=1alu[15].cout;
        q[]=1alu[].q;
        IF (q[15..8]==H"00") AND (q[7..0]==H"00") THEN
                zout=B"1";
        ELSE
                zout=B"0";
        END IF;
END;
```

We see from this design that the 1-bit ALU design file is included and instantiated as a component in a new design 16 times. Also, additional output signals are introduced to indicate output carry from the overall circuit and the value of the operation equals zero.

Data and Address Multiplexers

Typical circuits used in the data path are data and address multiplexers. As an example, consider the data bus multiplexer (DBUSMUX) used to provide the source data which will appear on the data bus lines. There are four possible sources of data on the input lines of this multiplexer as illustrated in Figure 6.9. They are the PC register, the TEMP register, the ALU, and the main memory.

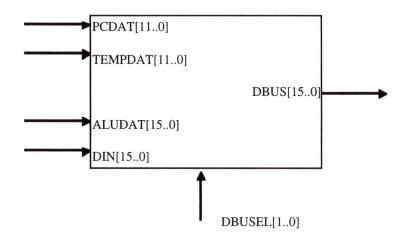

Figure 6.9 Data Bus Multiplexer

Two input lines DBUSEL[1..0] are used to select the source that is forwarded to the output of the multiplexer. Output lines represent the internal data bus lines. It should be noted that the PC and TEMP output lines are connected to the lower 12 bits of the data bus. If the contents of these registers is transferred to the data bus, the upper 4 bits will be grounded. This is clear in Example 6.6, the data bus multiplexer. Other multiplexers used in the data path are designed in a similar fashion.

Example 6.6 Data Bus Multiplexer.

```
TITLE "Data Bus Multiplexer dbusmux";

SUBDESIGN dbusmux
(
dbusel[1..0]                    :INPUT;
pcdat[11..0]                    :INPUT;
tempdat[11..0]                  :INPUT;
aludat[15..0]                   :INPUT;
din[15..0]                           :INPUT;
out[15..0]                           :OUTPUT;
)

VARIABLE
        pp[15..12]                   :NODE;

BEGIN
        pp[15..12]=GND;

        CASE dbusel[] IS

                WHEN B"00" =>
                out[11..0] = pcdat[];
                out[15..0] = pp[15..12];

                WHEN B"01" =>
                out[11..0] = tempdat[];
                out[15..12] = pp[15..12];

                WHEN B"10" =>
                out[] = aludat[];

                WHEN B"11" =>
                out[] = din[];
        END CASE;
END;
```

Data Path

The overall data path is integrated as the schematic (graphic) file just to visually describe connections of individual components designed using exclusively textual descriptions. It is described in a slightly simplified form in Figure 6.10. The dashed lines represent the control signals that are generated by the control unit to enable required register transfers or initiate local microoperations in registers. They also select source information which will be allowed to the bus.

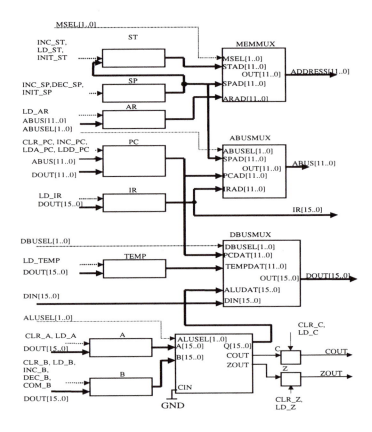

Figure 6.10 SimP's Data Path

The data path provides data input through external DIN[15..0] lines, data output through external DOUT[15..0] lines, addresses of memory locations or input/output registers through ADDRESS[11..0] lines, indications on the values of computation done in the arithmetic-logic unit through COUT and ZOUT lines, and finally, makes available instruction (placed in the instruction register IR) to the control unit so they can be decoded and executed. All registers of the data path are connected to the system clock and change values with the clock.

6.4.2 Control Unit Implementation

The control unit is the core of the SimP microprocessor. It provides proper timing and sequencing of all microoperations and perform the microoperations as required by the user program stored in the external memory. It provides proper start-up of the microprocessor upon power-up or manual reset. The control unit is also responsible for interrupt sequences as shown in preceding sections.

The global structure of the control unit is presented in Figure 6.11. It receives information from the data path both concerning the instructions that have to be executed and the results of ALU operations. It also accepts reset and interrupt request signals. Using these inputs it carries out the steps described in the control flowcharts of Figures 6.3 and 6.4. Obviously, in order to carry out proper steps in the appropriate machine (clock) cycles, a pulse distributor is needed to produce four periodic non-overlapping waveforms. They are used in conjunction with the information decoded by the operation decoder to determine actions, register transfers, and microoperations undertaken by the data path. The only exceptions occur in two cases presented by external hardware signals:

- When the system is powered-up or reset manually, the operation of the pulse distributor is seized for four machine cycles. This time is needed to initialize data path resources (program counter and stack pointer), as well as interrupt control circuitry.

- When the interrupt request signal is activated and if the interrupt structure is enabled, the control unit provides interruption of the current program upon the completion of the current instruction and the jump to predefined starting address of an interrupt service routine. The interruption is done by passing the control unit through the interrupt cycle.

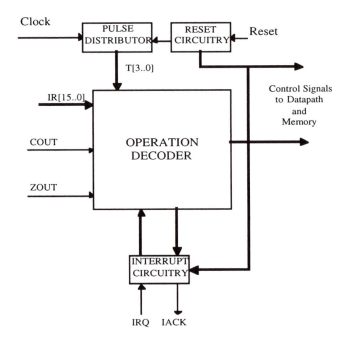

Figure 6.11 SimP's Control Unit

Pulse Distributor

The Pulse Distributor takes the system clock and provides four non-overlapping sequences called T[3..0]. The Pulse Distributor also has two input control lines as shown in Figure 6.12. The first, called "clear pulse distributor" (CLR), is used to bring the pulse distributor to its initial state T[3..0]=0001. It denotes that the T0 machine cycle is present. The second, called "enable pulse distributor" (ENA) is used to enable operation of the pulse distributor.

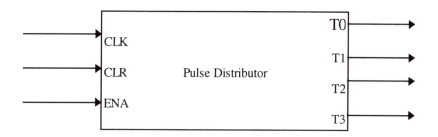

Figure 6.12 Pulse Distributor

The AHDL design file of the pulse distributor is given in Example 6.7.

Example 6.7 Pulse Distributor.

```
TITLE "Pulse Distributor pulsdist";

SUBDESIGN pulsdist
(
clk,clr,ena                   :INPUT;
t[3..0]                            :OUTPUT;
)

VARIABLE
ff[1..0]                   :DFF;

BEGIN
ff[].clk=clk;
ff[].clrn=!clr;

IF ena THEN
      ff[].d=ff[].q+1;
ELSE
      ff[].d=ff[].q;
END IF;

TABLE
      ff[1..0] => t3,t2,t1,t0;

      B"00"  =>     0,0,0,1;
      B"01"  =>     0,0,1,0;
      B"10"  =>     0,1,0,0;
      B"11"  =>     1,0,0,0;
```

```
END TABLE;
END;
```

Operation Decoder

The Operation Decoder represents the combinational circuit that recognizes input signals to the control unit, as well as the current state of the control unit in order to provide the proper control signals.

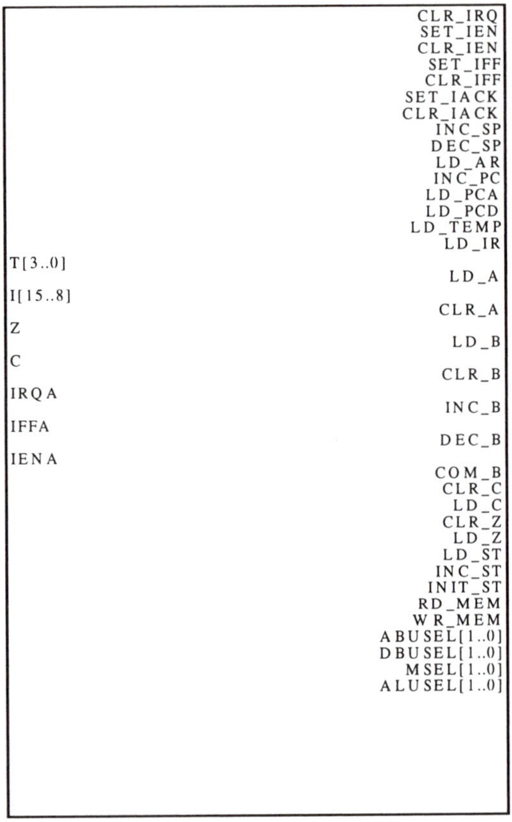

Figure 6.13 Operation Decoder

Input and output ports of the operation decoder are illustrated in Figure 6.13. Input ports are shown on the left hand side, and the control signals on the right hand side of the block representing the operation decoder. The AHDL design of the operation decoder is presented in Example 6.8.

Example 6.8 The Operation Decoder.

```
TITLE "Operation Decoder opdecode";

SUBDESIGN opdecode
(
t[3..0]                        :INPUT;
i[15..8]                       :INPUT;
z,c                            :INPUT;
irqa                           :INPUT;
iffa                           :INPUT;
iena                           :INPUT;
set_ien, clr_ien               :OUTPUT;
set_iff, clr_iff               :OUTPUT;
set_iack, clr_iack             :OUTPUT;
clr_irq                        :OUTPUT;
inc_sp, dec_sp                 :OUTPUT;
ld_ar                          :OUTPUT;
inc_pc                         :OUTPUT;
ld_pca, ld_pcd                 :OUTPUT;
ld_temp                        :OUTPUT;
ld_ir                                  :OUTPUT;
ld_a, clr_a                            :OUTPUT;
ld_b, clr_b, inc_b, dec_b, com_b       :OUTPUT;
clr_c, clr_z, ld_c, ld_z       :OUTPUT;
ld_st, inc_st, init_st         :OUTPUT;
rd_mem                         :OUTPUT;
wr_mem                         :OUTPUT;
abusel[1..0]                   :OUTPUT;% 1-sp,   2-pc,   3-ir %
dbusel[1..0]                   :OUTPUT;%0-pc,1-temp,2-alu,3-mem %
msel[1..0]                     :OUTPUT;% 0-sp,   1-ar 2-st %
alusel[1..0]                   :OUTPUT;% 0-add,1-and,2-a,3-b%
```

```
)

BEGIN

IF t[0] & !iffa THEN
        abusel[]= H"2";
        ld_ar=Vcc;
ELSIF t[0] & iffa THEN
        dbusel[]=B"00";
        msel[]=H"0";
        wr_mem=Vcc;
        clr_irq=Vcc;
END IF;

IF t[1] & !iffa THEN
        rd_mem=Vcc;
        msel[]=B"01";
        dbusel[]=B"11";
        ld_ir=Vcc;
        inc_pc=Vcc;
        ld_st=Vcc;
ELSIF t[1] & iffa THEN
        set_iack=Vcc;
        dec_sp=Vcc;
        init_st=Vcc;
END IF;
%start decoding of instruction%
IF t[2] & !iffa THEN
        IF i[15]==B"1" THEN
        %stack or funct. block addressing mode%
                abusel[]=B"01"; % ar<-sp %
                ld_ar=Vcc;
                dbusel[]=B"00"; % temp<-pc %
                ld_temp=Vcc;
                inc_st=Vcc;
        ELSIF i[15]==B"0" THEN
        %direct addressing mode%
                abusel[]=B"11"; % ar from ir12 %
                ld_ar=Vcc;
                dbusel[]=B"00"; % temp from pc %
                ld_temp=Vcc;
        END IF;
ELSIF t[2] & iffa THEN
```

```
                    msel[]=H"2";
                    dbusel[]=B"11";
                    rd_mem=Vcc;
                    ld_pcd=Vcc;
END IF;

IF t[3] & !iffa THEN

        CASE i[15..12] IS
        WHEN B"0000" =>
        %lda%
                ld_a=Vcc;
                rd_mem=Vcc;
                msel[]=B"01";
                dbusel[]=B"11";

        WHEN B"0001" =>
        %ldb%
                ld_b=Vcc;
                rd_mem=Vcc;
                msel[]=B"01";
                dbusel[]=B"11";

        WHEN B"0010" =>
        %sta%
                alusel[]=B"10";
                dbusel[]=B"10"; %from ALU%
                msel[]=B"01";
                wr_mem=Vcc;

        when B"0011" =>
        %stb%
                alusel[]=B"11";
                dbusel[]=B"10";
                msel[]=B"01";
                wr_mem=Vcc;

                WHEN B"0100" =>
        %jmp%
                abusel[]=B"11"; %from ir%
                ld_pca=Vcc;

        WHEN B"1000" =>
```

```
%jsr%
        dbusel[]=B"01";
        msel[]=H"1";
        wr_mem=Vcc;    % M[ar]<-temp %
        abusel[]=B"11";
        ld_pca=Vcc; % pc<-ir12 %
        dec_sp=Vcc;

WHEN B"1010" =>

        msel[]=H"1";
        wr_mem=Vcc;   % psha %
        dbusel[]=B"10";
        alusel[]=B"10"; %  M[ar]<-a %

WHEN B"1100" =>

        ld_a=Vcc;    % pula %
        msel[]=H"2";
        dbusel[]=B"11";
        rd_mem=Vcc;    % a<-M[st] %
        inc_sp=Vcc;

WHEN B"1110" =>
        msel[]=H"2";
        dbusel[]=B"11";          %ret%
        rd_mem=Vcc;
        ld_pcd=Vcc; % pc<-M[st] %
        inc_sp=Vcc;
END CASE;


CASE i[15..8] IS

WHEN H"71" =>

        ld_a=Vcc;       %add%
        dbusel[]=B"10"; %ALU select op%
        alusel[]=B"00";
        ld_c=Vcc;
        ld_z=Vcc;

WHEN H"72" =>
```

```
        ld_a=Vcc;       %aand%
        dbusel[]=B"10"; %ALU select op%
        alusel[]=B"01";

WHEN H"73" =>
        clr_a=Vcc;      %cla%

WHEN H"74" =>
        clr_b=Vcc;      %clb%

WHEN H"75" =>
        com_b=Vcc;      %cmb%

WHEN H"76" =>
        inc_b=Vcc;      %incb%

WHEN H"77" =>
        dec_b=Vcc;      %decb%

WHEN H"78" =>
        clr_c=Vcc;      %clc%

WHEN H"79" =>
        clr_z=Vcc;      %clz%

WHEN H"7A" =>
        set_ien=Vcc;

WHEN H"7B" =>
        clr_ien=Vcc;

WHEN H"7C" =>
IF c==B"1" THEN
        inc_pc=Vcc;     % sc %
ELSIF c==B"0" THEN
        abusel[]=B"11";
        ld_ar=Vcc;
END IF;

WHEN H"7D" =>

IF z==B"1" THEN
        inc_pc=Vcc;
```

```
        ELSIF z==B"0" THEN
                abusel[]=B"11";
                ld_ar=Vcc;
        END IF;

        END CASE;
        IF iena & irqa THEN
                set_iff=Vcc;
                clr_ien=Vcc;
        END IF;
ELSIF t[3] & iffa THEN
        clr_iack=Vcc;
        clr_iff=Vcc;
END IF;

END;
```

Reset Circuitry

Reset circuitry initializes the SimP at power-up or manual external reset. The only input is a reset signal, but several outputs are activated for proper initialization. The initialization consists of providing the initial values for the program counter and stack pointer, enabling the interrupt enable flip-flop (IEN), and clearing internal IFF flip-flops. Upon initialization, external interrupts are enabled and the control unit automatically enters the instruction execution cycle. This happens as soon as the pulse distributor is enabled and initialized. Reset circuitry is represented with its input and output ports in Figure 6.14.

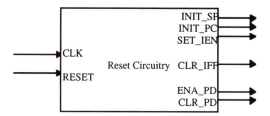

Figure 6.14 Reset Circuitry

Initialization lasts exactly four system clock cycles. In the case of an active RESET signal, an internal SR flip-flop is set. The output of the SR flip-flop represents the enable signal of an internal counter providing the internal counter will count until it reaches value 11. While this counter is counting, the initialization process is repeated in each machine cycle. When it stops, initialization is also stopped and the pulse distributor is enabled so it continues with its normal instruction execution cycle. The AHDL design file representing reset circuitry is given in Example 6.9.

Example 6.9 Initialization Circuitry.

```
TITLE "Initialization circuitry reset1";

SUBDESIGN reset1
(
        clk, reset              :INPUT;
        init_sp, clr_pc, set_ien, clr_iff, ena_pd, clr_pd
        :OUTPUT;
)

VARIABLE
        cnt[1..0]               :DFFE;
        cnt_out[1..0]           :NODE;
        rs                      :SRFF;
        ena_cnt                 :NODE;

BEGIN
        cnt_out[]=cnt[].q;
        rs.clk=clk;
        IF reset THEN
                rs.s=Vcc;
                rs.r=GND;
                ena_pd=GND;
                clr_pd=Vcc;
        ELSIF !reset OR cnt_out[]==B"11" THEN
                rs.r=Vcc;
                rs.s=GND;
                ena_pd=Vcc;
                clr_pd=GND;
        END IF;
        cnt[].clk=clk;
        ena_cnt=rs.q;
```

```
        IF ena_cnt THEN
                cnt[].d = cnt[].q + 1;
        ELSE
                cnt[].d = cnt[].q;
        END IF;
        IF ena_cnt THEN
                init_sp=Vcc;
                clr_pc=Vcc;
                set_ien=Vcc;
                clr_iff=Vcc;
        ELSIF !ena_cnt THEN
                init_sp=GND;
                clr_pc=GND;
                set_ien=GND;
                clr_iff=GND;
        END IF;
END;
```

Interrupt Circuitry

The interrupt circuitry has only one external input, the interrupt request (IRQ), and one external output, the interrupt acknowledgment (IACK). Upon interrupt request assertion (IRQA), an IRQ flip-flop is set producing the IRQA signal which is used by the operation decode circuitry. If the interrupts are enabled, and IRQA set, the operation decoder will set the interrupt flip-flop (IFF) to force the control unit to enter the interrupt cycle. In the interrupt cycle, the IACK flip-flop, whose output is available to circuitry outside the SimP, is set for two machine cycles. The interrupt enable flip-flop (IEN) can be set by the operation decoder or reset circuitry and the interrupt flip-flop can be cleared by the operation decoder or reset circuitry. The AHDL file describing the operation of the interrupt circuitry is given in Example 6.10

Example 6.10 Interrupt Circuitry.

```
TITLE "Interrupt circuitry interrupt";
SUBDESIGN interrupt
(
set_ien, set_ien1, clr_ien, set_iff, clr_iff, clr_iff1    :INPUT;
set_iack, clr_iack, irq, clk                   :INPUT;
iffa, irqa, iena, iack                         :OUTPUT;
)
VARIABLE
```

```
irqff, ienff, iackff, iff                    :DFF;
clr_irq                                      :NODE;
BEGIN
clr_irq=iackff.q;
irqff.clk=clk;
iackff.clk=clk;
ienff.clk=clk;
iff.clk=clk;

IF set_ien # set_ien1 THEN
        ienff.d=Vcc;
ELSIF clr_ien THEN
        ienff.d=GND;
ELSE
        ienff.d=ienff.q;
END IF;
IF set_iff   THEN
        iff.d=Vcc;
ELSIF clr_iff # clr_iff1 THEN
        iff.d=GND;
ELSE
        iff.d=iff.q;
END IF;
IF set_iack THEN
        iackff.d=Vcc;
ELSIF clr_iack THEN
        iackff.d=GND;
ELSE
        iackff.d=iackff.q;
END IF;
IF irq THEN
        irqff.d=Vcc;
ELSIF clr_irq THEN
        irqff.d=GND;
ELSE
        irqff.d=irqff.q;
END IF;
iack=iackff.q;
irqa=irqff.q;
iena=ienff.q;
iffa=iff.q;
END;
```

Control Unit

The overall control unit circuit is represented by the schematic diagram in
Figure 6.15.

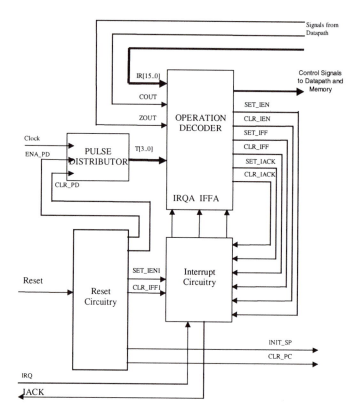

Figure 6.15 Control Unit

6.4.3 Design Analysis

The SimP design, as presented in this chapter is accomplished in a hierarchical
manner. No attempt at design optimization has been made. After designing
components of the bottom level, the next level of hierarchy is designed. The
final integration of the data path and control unit was done using schematic

entry because it is represented by the pure structural design. Data summarizing the final compilation process are shown in Table 6.6.

Table 6.5 Compilation Times.

Compilation Times	
Compiler Nettles Extractor	00:00:07
Database Builder	00:00:43
Logic Synthesizer	00:00:38
Partitioner	00:00:25
Fitter	00:00:56
Timing SNF Extractor	00:00:41
Assembler	00:00:04
Total Time	00:03:34

The Max+PLUS II compiler was allowed to perform all resource assignments automatically. The device used to fit the design was selected automatically from the FLEX 8000 series. The compiler found that the device suitable for SimP implementation is the EPF8820 with the basic features given in Table 6.7.

Table 6.6 SimP Features.

Input pins	19
Output pins	33
Logic Cells required	541
Utilization of logic cells	541/672 = 80%
Total flip-flops required	117
Total dedicated pins used	4

It should be noticed that the data bus in the compiled design is separated into input and output data lines.

Extensive timing simulation was performed on each hierarchical level of design. It showed compliance with the system requirements. The maximum clock frequency achieved at the simulation level reached 16.6 MHz or a basic clock period (machine cycle) of 60 ns.

The SimP's design gives various opportunities for performance improvements and are left to the readers as exercises in using the design tools and FPLD technology.

7 DESIGNING WITH ADVANCED TOOLS - AN INTRODUCTION TO VHDL

VHDL is the VHSIC (Very High Speed Integrated Circuits) Hardware Description Language used to develop, document, simulate, and synthesize the design of electronic systems and chips. This chapter provides an introduction to VHDL as a more abstract and powerful hardware description language which is accepted as an IEEE standard. The goal of this chapter is to demonstrate how VHDL can be used in digital system design. It should help hardware designers of FPLD- based systems to better model their designs. A subset of the language features is used to provide designs that can almost always be synthesized. This chapter is also intended to bring a novice designer to the level of writing fairly complex VHDL descriptions. We will use the IEEE standard VHDL to provide compatibility with other models and to show some specifics of Altera's implementation. Combined with the flexibility and potential reconfigurability of FPLDs, VHDL represents a tool which will be increasingly used in digital system design and prototyping.

7.1. Overview of VHDL

VHDL is an industry standard hardware description language used to document electronic systems design from an abstract to a concrete level. VHDL is also used to standardize inputs and outputs from various CAE tools including simulation tools, synthesis tools, and layout tools. It will be shown how to write efficient and correct VHDL descriptions of hardware designs at different levels of abstraction. These levels of abstraction range from the initial specification to the gate-level implementation. All examples in this chapter are described in the IEEE standard 1076- 1987 VHDL and are compiled and simulated with the Altera Max+PLUS II compiler and simulation tools. The best way to approach VHDL is to initially learn a subset of the language, and learn the more complex models and features as they are required.

VHDL consists of several parts organized as follows:

- The actual VHDL language

- Additional data type declarations in the Package **STANDARD**, and some utility functions in the Package **TEXTIO**; these two packages form a Library called **STD** (A combination of the **TEXTIO** and **STANDARD** packages)

- A **WORK** library reserved for user's designs

- A vendor package with vendor libraries

- User packages and libraries

A VHDL description lists a design's components, interconnections, and documents the system behavior. A VHDL description can be written at various levels of abstraction:

- algorithmic

- register transfer

- gate level functional with unit delay

- gate level with detailed timing

Using a top-down design methodology, a designer first represents an abstract system and latter develops it into a more detailed description. Some design decisions can be left for the latter phases of design. VHDL provides ways of abstracting a design or "hiding" implementation details. A designer can design with top-down successive refinements specifying more details of how the design is built.

A design description or model, written in VHDL, can be run through a VHDL simulator to demonstrate its behavior. Simulating design models requires a simulation stimulus to provide a way of observing the model during simulation, and for capturing the results of simulation for later inspection. VHDL supports a variety of data types useful to the hardware modeler for both simulation and synthesis, as well as bits, Booleans, and numbers, which are defined in the Package **STANDARD**.

Some parts of VHDL can be used with logic synthesis tools for producing a physical design. Many VLSI gate-array vendors can convert a VHDL design description into a gate level netlist to build a customized integrated circuit component. Hence the applications of VHDL are:

• Documenting a design

• Simulating behavior

• Synthesizing logic

Designing in VHDL is like programming in many ways. Compiling and running a VHDL design is similar to compiling and running other programming languages. The result of compiling an object module is produced and placed in a special VHDL library. Simulation is done subsequently by selecting the object units from the library and loading them in to the simulator. The main difference is that a VHDL design always runs in simulated time and events occur in successive time steps.

There are several differences between VHDL and conventional programming languages. The major differences are the notion of delay and simulation environment. Concurrency and component netlisting also are not found in conventional programming languages. VHDL supports concurrency using the concept of concurrent statements running in simulated time. Simulated time is a feature found only in simulation languages. Also, there are sequential statements in VHDL to describe algorithmic behavior.

Design hierarchy in VHDL is accomplished by separately compiling components that are instanced in a higher-level component. The linking process is done either by the compiler or the simulator using the VHDL library mechanism.

Some software systems have version control systems to generate different versions of the program. VHDL has a configuration capability for generating design variations.

7.1.1 VHDL Designs

A VHDL design consists of several separate design units, each of which is compiled and saved in a library. The four design units that can be compiled are:

The **Entity unit** describes the interface signals and represents the most basic building block in a design. If the design is hierarchical, the top-level description (entity) will have lower-level descriptions (entities)contained in it.

The **Architecture unit** describes behavior. A single entity can have multiple architectures, behavioral or structural for example.

The **Configuration unit** selects a variation of a design from a design library. It is used to bind a component instance to an entity-architecture pair. A configuration can be considered as a parts list for a design. It describes which behavior to use for each entity.

The **Package unit** stores together, for convenience, certain frequently used specifications such as data types and subprograms used in a design. The Package unit can be considered a toolbox used to build designs.

Typically, an architecture uses previously compiled components from an ASIC vendor library. Once compiled, a design becomes a component in a library that may be used in other designs. Additional compiled vendor's packages are also stored in the library.

By separating the entity (I/O interface of a design) from its actual architecture implementation, a designer can change one part of a design without recompiling other parts. In this way a feature of reusability is implemented. For example, a CPU containing a precompiled ALU saves recompiling time. Configurations, named and compiled units stored in the library, provide an extra degree of flexibility by saving variations of a design (for example, two versions of CPU, each with a different ALU).

The designer defines the basic building blocks of VHDL in the following sections:

- Library
- Package
- Entity
- Architecture
- Configuration

An example of a design unit is shown in Example 7.1.

Example 7.1 mine.vhd.

```
PACKAGE my_units IS                    --Package--
CONSTANT unit_delay:   TIME :=1 ns;
END my_units

ENTITY compare IS                 --Entity--
PORT  (a, b : in bit ;
        c : out bit);
END compare

LIBRARY my_library;                --Architecture--
USE my_library.my_units.all;

ARCHITECTURE first OF compare IS
BEGIN
c <=NOT (a XOR b) after unit_delay;
END first
```

There are three design units in design mine.vhd. After compilation, there are four compiled units in the library my_library:

- Package my_units - provides a sharable constant

- Entity compare - names the design and signal ports

- Architecture first of compare - provides details of the design

- A configuration of compare - designates first as the latest compiled architecture

Each design unit could be in a separate file and could be compiled separately, but the other of compilations must be as shown in Example 7.1. The package my_units can also be used in other designs. The design entity compare can now be accessed for simulation or used as a component in another design. To use compare, two input values of type bit are required at pins a and b ; 1 ns latter a '1' or '0' appears at output pin c. Basic VHDL design units are described in more details in the following sections.

Keywords of the language and types provided with the STANDARD package are given and will be shown in capital letters. For instance, in Example 7.1, the keywords are ARCHITECTURE, PACKAGE, ENTITY, BEGIN, END, IS. Names of user-created objects, such as compare, will be shown in lowercase letters. However, it should be pointed out, VHDL is not case sensitive and this convention is used just for readability.

7.1.2 Library

The results of a VHDL compilation are stored in a library for subsequent simulation or for use in further or other designs. A library can contain:

- A package (shared declarations)

- An entity (shared designs)

- An architecture (shared design implementations)

- A configuration (shared design versions)

The two built-in libraries are WORK and STANDARD, but the user can create other libraries. VHDL source design units are compiled into the WORK library unless a user directs it to another library.

To access an existing library unit in a library as a part of a new VHDL design, the library name must be declared first. The syntax is:

```
LIBRARY logical_name;
```

Now, component designs compiled into the specified library can be used. Packages in the library can be accessed by way of a subsequent USE statement. If the WORK library is used, it does not need to be declared.

Compiled units within a library can be accessed with up to three levels of names:

```
<library_name>.<package_name>.<item_name>

<library_name>.<item_name>

<item_name>
```

The last from is used if the WORK library is assumed.

Units in a library must have unique names; all design entity names and package names are unique within a library. Architecture names need to be unique to a particular design entity.

In order to locate a VHDL library in a file system, it is sometimes necessary to issue the commands outside of the VHDL language. This is compiler and system dependent and a user must refer to the appropriate vendor manuals for the commands.

7.1.3 Package

The next level of hierarchy within a library is a package. A package collects a group of related declarations together. Typically, a package is used for:

• Subprogram declarations

- Type declarations

- Constant declarations

- File declarations

- Alias declarations

A Package is created to store common subprograms, data types, constants and compiled design interfaces that will be used in more than one design. This strategy promotes reusability. A package consists of two separate design units: the package header (which identifies all of the names and items) and the optional package body (which gives more details of the named item).

All vendors provide a package named STANDARD in a predefined library named STD. This package defines useful data types, such as BIT, BOOLEAN, and BIT_VECTOR. There is also a text I/O package called TEXTIO in STD.

A USE clause allows access to a package in a library. No USE clause is required for the Package STANDARD. The default is:

```
LIBRARY STD;

USE STD.STANDARD.ALL:
```

Additionally, component or CAD tool vendors provide packages of utility routines and design pieces to assist design work. For example, VHDL descriptions of frequently used CMOS gate components are compiled into a separate library and their declarations are kept in a package.

7.1.4 Entity

The design entity defines a new component name, its input/output connections, and related declarations. The entity represents the I/O interface to a component design. VHDL separates the interface to a design from the details of architectural implementation. The entity describes the type and direction of signal connections while an architecture describes the behavior of a component. After an entity is compiled into a library, it can be simulated or used as a component in another design. An entity must have a unique name within a library. If a component has signal ports, they are declared in an entity declaration. The syntax used to declare an entity is shown in Example 7.2.

Example 7.2 Entity Syntax.

```
ENTITY entity_name IS

        [generics]    -- [  ] designate optional part
        [ports]
        [declarations  {constants, types, signals}]
        -- {  } designate one or more instances
        [begin         statements] --Typically not used

END entity_name;
```

An entity specifies external connections of a component. In the Figure 7.1 , an AND gate (andgate) with two input signals and one output signal is shown.

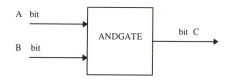

Figure 7.1 Example of AND gate

Figure 7.1 emphasizes the interface to the design. All signals are bit type which mandates their usage; the andgate design only works on bit type data. VHDL declaration of this entity is shown in Example 7.3.

Example 7.3 VHDL AND Gate Entity.

```
        ENTITY andgate IS
        PORT (a, b: IN BIT;
              c: OUT BIT);
        END andgate;
```

In Example 7.3, andgate is defined as a new component. The reserved word IS is followed by the port declarations with their names and types.

Any declaration used in an entity port must be previously declared. When an entity is compiled into a library, it becomes a component design that can be used in another design. A component can be used without knowledge of its internal design details.

All designs are created from entities. An entity in VHDL corresponds directly to a symbol in the traditional schematic entry methodology. The input ports in the Example 7.3 directly correspond to the two input pins and the output port corresponds to the output pin.

7.1.5 Architecture

An architecture design unit specifies the behavior, interconnections, and components of a previously compiled design entity. The architecture defines the function of design entity. It specifies the relationships between the inputs and outputs that must be expressed in terms of behavior, dataflow, or structure. The Entity design unit must be compiled before the compilation of its architecture. If an entity is recompiled, all its architectures must also be recompiled.

VHDL allows a designer to model a design at several levels of abstraction or with various implementations. An entity may be implemented with more than one architecture. Figure 7.2 illustrates two different architectures of the entity andgate.

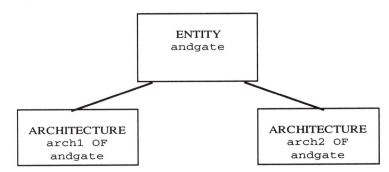

Figure 7.2 Entity with Two Different Architectures

All architectures have identical interfaces, but each needs a unique architecture name. A designer selects a particular architecture of a design entity during configuration (for example `arch1`).

VHDL architectures are generally categorized in styles as:

- Behavioral (defines sequentially described process)
- Dataflow (implies a structure and behavior)
- Structural (defines interconnections of components)

A design can use any or all of these design styles. Generally, designs are created hierarchically using previously compiled design entities. They can only be combined using a structural style which looks like a list of components wired together, such as a netlist. The architecture is defined in VHDL with the syntax of Example 7.4.

Example 7.4 Architectural Syntax.

```
ARCHITECTURE architecture_name OF entity_name IS

        [architecture_declarative_part]

BEGIN

        [architecture_statement_part]

END [architecture_name];
```

The `architecture_declarative_part` is where items used only in this architecture such as types, subprograms, constants, local signals and components are declared. The `architecture_statement_part` is the actual design description. All statements between the `BEGIN` and `END` statement are called concurrent statements because all statements execute concurrently.

The architecture can be considered as a counterpart of the schematic for the component in a traditional design.

7.1.6 Behavioral Style Architecture

An example of the architecture called `arch1` of entity `andgate` is shown in Example 7.5.

Example 7.5 Behavioral Style Architecture.

```
ARCHITECTURE arch1 OF andgate IS
BEGIN
        PROCESS (a, b);
        BEGIN
                IF a ='1' AND b ='1' THEN
                c <='1' AFTER 1 ns;
                ELSE
                c <='0' AFTER 1 ns;
                ENDIF;
        END PROCESS;
    END arch1;
```

It contains a process that uses signal assignment statements. If both input signals a and b have the value '1', c gets a '1'; otherwise c gets a '0'. This architecture describes a behavior in a "program-like" or algorithmic manner.

VHDL processes may run concurrently. The list of signals for which the process is waiting (sensitive to) is shown in parentheses after the word process. Processes wait for changes in an incoming signal. A process is activated whenever input signals change. The output delay of signal c depends on the AFTER clause in the assignment.

Parallel operations can be represented with multiple processes. An example of processes running in parallel is shown in Figure 7.3. The processes communicate with each other getting their data from an outside signal. Inside, the process operates with variables. The variables are local storage and cannot be used to transfer information outside the process. Sequential statements, contained in process, execute in order of appearance as in conventional programming languages.

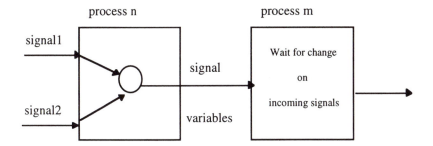

Figure 7.3 Process model

Process m in Figure 7.3 receives signals from process n. The running of one process can depend upon results of operation of another process.

In a top-down design style, a behavioral description is usually the first step; the designer focuses on the abstract design behavior. Later, the designer can choose the precise signal-bus and coding.

7.1.7 Dataflow Style Architecture

A dataflow architecture models the information or dataflow behavior of combinational logic functions such as adders, comparators, multiplexers, decoders, and other primitive logic circuits. The Example 7.6 defines the entity and architecture, in a dataflow style, of xor2, an exclusive-OR gate. xor2 has input ports a and b of type BIT, and an output port c of type BIT. There is also a delay parameter m, which defaults to 1.0 ns. The architecture dataflow gives output c, the exclusive-OR of a and b after m (1ns).

Example 7.6 XOR2

```
ENTITY xor2 IS
        GENERIC (m: time :=1.0 ns);
        PORT (a, b: IN BIT;
            c: OUT BIT);
END xor2
```

```
ARCHITECTURE dataflow OF xor2 IS
BEGIN
        c <= a XOR b AFTER m;
END dataflow;
```

Once this simple gate is compiled into a library, it can be used as a component in another design by referring to the entity name (xor2) and providing three port parameters and an optional delay parameter.

7.1.8 Structural Style Architecture

Top-level VHDL designs use structural style to instance and connect previously compiled designs. The Example 7.7 uses two gates xor2 (exclusive-OR) and inv (inverter) to realize a simple comparator. The schematic in Figure 7.4 represents a comparator.

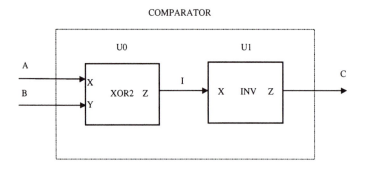

Figure 7.4 Schematic representation of comparator

Inputs in the circuit, labeled a and b, are inputs into first xor2 gate. The signal wire I from xor2 connects to the next component inv, which provides an output c.

Example 7.7 Comparator.

```
ENTITY comparator IS
        PORT(a, b: IN BIT; c: OUT BIT);
END comparator;

ARCHITECTURE structural OF COMPARATOR IS
SIGNAL i: BIT;
COMPONENT xor2 PORT (x,y: IN BIT; z: OUT BIT);
END COMPONENT;

COMPONENT inv PORT (x: IN BIT; z: OUT BIT);
END COMPONENT;

BEGIN
        u0:    xor2 PORT MAP (a, b, i);
        u1:    inv PORT MAP (i, c);
END structural;
```

The architecture has an arbitrary name structural. Local signal i is declared in the declaration part of architecture. Component declarations are required unless these declarations are placed in a package. Two components are given instance names (u0 and u1). The port map indicates the signal connections to be used. The design entities xor2 and inv are found in library WORK, since no library is declared.

Which architecture will be used depends on the accuracy wanted and whether structural information is required. If the model used for PCB layout purposes, then the structural architecture is probably most appropriate. For simulation purposes, behavioral models are probably more efficient in terms of the memory space and speed of execution.

7.1.9 Configuration

The configuration assists the designer in experimenting with different design by selecting particular architectures. Two different architectures of the entity andgate, called arch1 and arch2, are illustrated in Figure 7.2. A configuration selects a particular architecture, for example arch1, from a library. The syntax is shown in Example 7.8.

Example 7.8 Configuration Syntax.

```
CONFIGURATION identifier OF entity_name IS

        specification

END identifier;
```

Different architectures may use different algorithms or levels of abstraction. If the design uses a particular architecture, a CONFIGURATION statement is used. A configuration is a named and compiled unit stored in the library. The VHDL source description of a configuration identifies, by name, other units from a library. This is shown in Example 7.9.

Example 7.9 VHDL Source Configuration Description.

```
CONFIGURATION alu1_fast OF alu IS

FOR alu1

FOR u0: comparator USE ENTITY WORK.comparator(dataflow);
```

In Example 7.9, configuration alu1_fast is created for the entity alu and architecture alu1. The USE clause identifies a library, entity, and architecture of the component comparator. The final result is the configuration called alu1_fast. It is a variation of the design alu. Configuration statements permit selection of a particular architecture. When no explicit configuration exists, the latest compiled architecture is used (it is called a null configuration).

The power of the configuration is that recompilation of the entire design is not needed when using another architecture; instead, only recompilation of the new configuration is needed.

7.2 Specifics of VHDL in Max+PLUS II Design Environment and VHDL Synthesis

VHDL is fully integrated into the Max+PLUS II design environment. VHDL designs can be entered with any text editor and compiled to create output files for simulation, timing analysis, and device programming. Max+PLUS II supports a subset of IEEE 1076-1987 VHDL and is described in Altera's documentation. VHDL design files (.vhd) can be combined with other design files into a hierarchical design called a project. Other types of files include Altera specific AHDL design files (TDF files), schematic entry files (GDF files), Waveform Design files (WDF files), Altera design files (ADF files), and State Machine files (SMF files), as well as industry standard EDIF files or Xilinx Netlist format files (XNF files). Each file in a project hierarchy is connected through its input and output ports to one or more design files at the next higher hierarchy level.

The Max+PLUS II environment allows a designer to create a symbol that represents a VHDL design file and incorporate it into graphic design file. Custom functions, as well as Altera-provided macrofunctions, can be incorporated into any VHDL design file. The Max+PLUS II Compiler automatically processes VHDL design files and optimizes them for Altera FPLD devices. The compiler can create a VHDL Output File (.vho) that can be imported into an industry standard environment for simulation. Or, after a VHDL project has compiled successfully, optional simulation and timing analysis with Max+PLUS II can be done, and an Altera device can be programmed.

Altera provides the `altera` library that includes the `maxplus2` package, which contains all Max+PLUS II primitives and macrofunctions supported by VHDL. Besides that Altera provides several other packages located in subdirectories of the \maxplus2\-max2vhdl directory. They are shown in Table 7.1.

Table 7.1 Max+PLUS II Packages

Package	Library	Contents
maxplus2	altera	Max+PLUS II primitives and macrofunctions supported by VHDL
std_logic_1164	ieee	Standard for describing interconnection data types for VHDL modeling, and the STD_LOGIC and STD_LOGIC_VECTOR types
std_logic_arith	ieee	SIGNED and UNSIGNED types, arithmetic and comparison functions for use with SIGNED and UNSIGNED types, and the conversion functions CONV_INTEGER, CONV_SIGNED, and CONV_UNSIGNED
std_logic_signed	ieee	Functions that allow Max+PLUS II to use STD_LOGIC_VECTOR types as if they are SIGNED types
std_logic_unsigned	ieee	Functions that allow Max+PLUS II to use STD_LOGIC_VECTOR types as if they are UNSIGNED types

Altera also provides the STD library with the STANDARD and TEXTIO packages that are defined in the IEEE standard VHDL Language Reference Manual. This library is located in the \maxplus2\max2vhdl\std directory.

7.2.1 Combinatorial Logic Implementation

Combinatorial logic is implemented in VHDL with Concurrent Signal Assignment Statements or with Process Statements that describe purely combinatorial behavior. Both of these statements should be placed in the Architecture Body of a VHDL design File, as shown in Example 7.10.

Example 7.10 VHDL Design Architecture Body

```
ARCHITECTURE arch_name OF and_gate IS
BEGIN
        [Concurrent_Signal_Assignments]
        [Process_Statements]
        [Other Concurrent Statements]
END arch_name;
```

Concurrent Signal Assignment Statements assign values to signals, directing the Compiler to create simple gates and logical connections. Three basic types of Concurrent Signal Assignment Statements are used:

- Simple Signal Assignment

- Conditional Signal Assignment

- Selected Signal Assignment

The simple Signal Assignment Statements shown in Example 7.11 create an AND gate and connect two nodes, respectively. These statements are executed concurrently.

Example 7.11 Signal Assignment Statements.

```
ENTITY simpsig IS
      PORT (a, b: IN BIT;
              c, d: OUT BIT);
END simpsig;

ARCHITECTURE arch1 OF simpsig IS
BEGIN
        c <= a AND b;
        d <= b;
```

```
END arch1;
```

Figure 7.5 shows a Graphic Design File that is equivalent to the simple Signal Assignment Statements of Example 7.11.

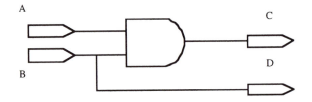

Figure 7.5 GDF equivalent of Simple Signal Assignment Statements

Conditional Signal Assignment Statements list a series of expressions that are assigned to a target signal after the positive evaluation of one or more Boolean expressions. Example 7.12 shows a basic 2-to-1 multiplexer in which the input value i0 is assigned to output y when s equals '0'; otherwise, the value i1 is assigned to output y.

Example 7.12 2-to-1 Multiplexer.

```
ENTITY condsig IS
        PORT (i0, i1, s: IN BIT;
              y: OUT BIT);
END condsig;

ARCHITECTURE arch1 OF condsig IS
BEGIN
        y <= i0 WHEN s = '0' ELSE i1;
END arch1;
```

The GDF equivalent of this example is shown in Figure 7.6. When a Conditional Signal Assignment Statement is executed, each Boolean expression is tested in the order in which it is written. The value of the expression preceding the WHEN keyword is assigned to the target signal for the first Boolean expression that is evaluated as TRUE. If none of the Boolean expressions are TRUE, the expression following the last ELSE keyword is assigned to the target.

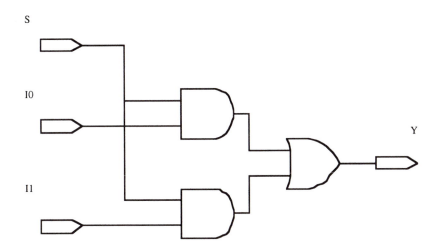

Figure 7.6 GDF Equivalent of Conditional Assignment Statement with One Alternative

An example of a priority encoder is shown in Example 7.13.

Example 7.13 Priority Encoder.

```
ENTITY condsig IS
        PORT (high, mid, low: IN BIT;
                q: OUT INTEGER);
END condsig;

ARCHITECTURE arch1 OF condsig IS
BEGIN
        q <=    3 WHEN high = '1' ELSE
                2 WHEN mid = '1' ELSE
                1 WHEN low = '1' ELSE
                0;
END arch1;
```

Figure 7.7 shows the GDF equivalent of priority encoder as synthesized by the Compiler. Selected Signal Assignment Statement lists alternatives that are available for each value of an expression, then selects a course of action based on the value of expression.

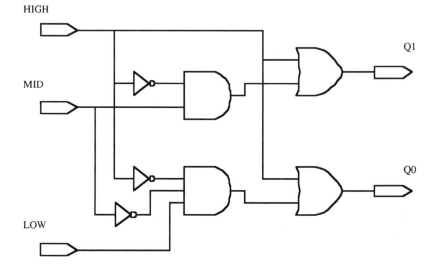

Figure 7.7 GDF equivalent of priority encoder

In Example 7.14 the Selected Signal Assignment Statement is used to create multiplexer.

Example 7.14 Selected Signal Assignment Statement.

```
ENTITY selsig IS
        PORT ( d0, d1, d2, d3: IN BIT;
               s: IN INTEGER RANGE 0 TO 3;
               y: OUT BIT);
END selsig:
```

```
ARCHITECTURE arch1 OF selsig IS
BEGIN

WITH s SELECT
        y <=    d0 WHEN 0,
                d1 WHEN 1,
                d2 WHEN 2,
                d3 WHEN 3;

END arch1;
```

The select expression, found between the WITH and SELECT keywords, selects the signal to be assigned to the target signal. Figure 7.8 shows the GDF equivalent to the Selected Signal Assignment Statement of Example 7.14.

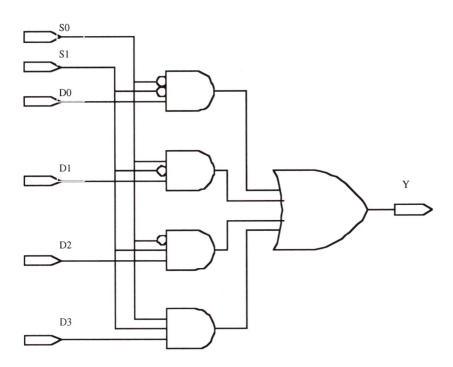

Figure 7.8 GDF equivalent of 4-to-1 multiplexer

Process statements include a set of sequential statements that assign values to signals. Process statements that describe purely combinatorial behavior can be used to create combinatorial logic. To ensure that a process is combinatorial, its sensitivity list must contain all signals that are read in the process. Example 7.15 is an example of the process that counts the number of bits with the value 1 in a 4-bit word represented by the signal d.

Example 7.15 Counter.


```
ENTITY counter IS
        PORT ( d: IN BIT_VECTOR (3 DOWNTO 0);
                 q: OUT INTEGER RANGE 0 TO 4);
END counter;

ARCHITECTURE arch1 OF counter IS
BEGIN
        PROCESS (d)
                VARABLE n: INTEGER;
        BEGIN
                n := 0;

                FOR i IN d'RANGE LOOP
                     IF d(i) = '1' THEN
                        n = n + 1;
                     END IF;
                END LOOP;

                q <= n;
        END PROCESS;

END arch1;
```

The signal d is the only signal contained in the sensitivity list that follows the PROCESS keyword. This signal is declared as an array in the entity declaration. If d(i) equals 1, the IF statement increments the internal counting variable n. The n variable is then assigned to the signal q, which is also declared in the Entity Declaration.

7.2.2 Sequential Logic Implementation

Sequential logic is implemented in VHDL with process statements. Process statements direct the compiler to create logic circuitry that is controlled by the process statement's Clock signal.

Registers and Counters Synthesis

A register is implemented implicitly with a Register Inference. Register Inferences in Max+PLUS II VHDL support any combination of Clear, Preset, Clock Enable, and asynchronous Load signals. The Compiler can infer memory elements from the following VHDL statements which are used within a process Statement:

IF statements can be used to imply registers for signals and variables in the clauses of the IF statement while WAIT statements can be used to imply registers in a synthesized circuit. The Compiler creates flip-flops for all signals and some variables that are assigned values in any process with a WAIT statement. The WAIT statement must be listed at the beginning of the process statement.

Registers can be also implemented with the Component Instantiation Statement. However, Register Inferences are technology independent.

Example 7.16 shows several ways to infer registers that are controlled by a Clock and asynchronous Clear, Preset, and Load signals:

Example 7.16 Register Inference.

```
ENTITY register_inference IS
        PORT ( d, clk, clr, pre, load, data: IN BIT;
               q1, q2, q3, q4, q5: OUT BIT);
END register_inference;

ARCHITECTURE arch1 OF register_inference IS

BEGIN

        -- Register with active-low clock
```

```
       PROCESS
       BEGIN
              WAIT UNTIL clk = '0';
              q1 <= d;
       END PROCESS;

       -- Register with active-high clock and  asynchronous
clear
       PROCESS
       BEGIN
              IF clr = '1' THEN
                     q2 <= '0';
              ELSIF clk'EVENT AND clk = '1' THEN
                     q2 <= d;
              END IF
       END PROCESS;

       -- register with active-high clock and asynchronous
preset
       PROCESS
       BEGIN
              IF pre = '1' THEN
                     q3 <= '1';
              ELSIF clk'EVENT AND clk = '1' THEN
                     q3 <= d;
              END IF;
       END PROCESS;

       -- Register with active-high clock and asynchronous load
       PROCESS
       BEGIN
              If load = '1' THEN
                     q4 <= data;
              ELSIF clk'EVENT AND clk = '1' THEN
                     q4 <= d;
              END IF;
       END PROCESS;

       -- Register with active-low clock and asynchronous clear
and preset
       PROCESS
       BEGIN
              If clr = '1' THEN
```

```
                        q5 <= '0';
              ELSIF pre = '1' THEN
                        q5 <= '1';
              ELSIF clk'EVENT AND clk = '0' THEN
                        q5 <= d;
         END PROCESS;
END arch1;
```

All the processes in Example 7.16 are only sensitive to changes to the control signals (clk, clr, pre, and load) and to changes of the data signal data.

A counter can be implemented with a Register Inference. A counter is inferred from an IF statement that specifies a Clock edge together with logic that adds or subtracts a value from the signal or variable. The IF statement and additional logic should be inside a Process Statement. Example 7.17 shows several 8-bit counters controlled by the clk, clear, ld, d, enable, and up_down signals that are implemented with IF statements.

Example 7.17 8-bit Counters Implemented with IF Statements.

```
ENTITY counters IS
        PORT (d : IN INTEGER RANGE 0 TO 255;
                  clk, clear, ld, enable, up_down: IN BIT;
                  qa, qb, qc, qd, qe, qf: OUT INTEGER
                  RANGE 0 TO 255);
END counters;

ARCHITECTURE arch OF counters IS
BEGIN


        -- An enable counter

        PROCESS (clk)
                VARIABLE cnt: INTEGER RANGE 0 TO 255;
        BEGIN
                IF (clk'EVENT AND clk = '1') THEN
                        IF enable = '1' THEN
                                cnt := cnt + 1;
                        END IF;
```

```
        END IF;

        qa <= cnt;
END PROCESS;


-- A synchronous load counter

PROCESS (clk)
        VARIABLE cnt: INTEGER RANGE 0 TO 255;
        IF   (clk'EVENT   AND    clk   =   '1')   THEN
            IF ld = '0' THEN
                    cnt := d;
            ELSE
                    cnt := cnt +1;
            END IF;
        END IF;

        qb <= cnt;
END PROCESS;


-- An up_down counter

PROCESS (clk)
        VARIABLE cnt: INTEGER RANGE 0 TO 255;
        VARIABLE direction: INTEGER;
BEGIN
        IF (up_down = '1') THEN
            direction := 1;
        ELSE
            direction := -1;
        END IF;

        IF (clk'EVENT AND clk ='1') THEN
            cnt := cnt + direction;
        END IF;

        qc <= cnt;
END PROCESS;


-- A synchronous clear counter
```

```
PROCESS (clk)
        VARIABLE cnt: INTEGER RANGE 0 TO 255;
BEGIN
        IF (clk'EVENT AND clk = '1') THEN
                IF clear = '0' THEN
                        cnt := 0;
                ELSE
                        cnt := cnt + 1;
                END IF;
        END IF;

        qd <= cnt;
END PROCESS;


-- A synchronous load clear counter

PROCESS (clk)
BEGIN
        IF (clk'EVENT AND clk = '1') THEN
                IF clear = '0' THEN
                        cnt := 0;
                ELSE
                        IF ld = '0' THEN
                                cnt := d;
                        ELSE
                                cnt := cnt +1;
                END IF;
        END IF;

        qe <= cnt;
END PROCESS;


-- A synchronous load enable up_down counter

PROCESS (clk)
        VARIABLE cnt: INTEGER RANGE 0 TO 255;
        VARIABLE direction: INTEGER;
BEGIN
        IF up_down = '1' THEN
                direction := 1;
```

```
                ELSE
                        direction := -1;
                END IF;

                IF (clk'EVENT AND clk = '1') THEN
                        IF ld = '0' THEN
                                cnt := d;
                        ELSE
                                IF enable = '1' THEN
                                        cnt := cnt + direction:
                                END IF;
                        END IF;
                END IF;

                gf <= cnt;
        END PROCESS;
END arch;
```

All processes in Example 8.17 are only sensitive to changes on the `clk` input signal. All other control signals are synchronous. At each Clock edge, the `cnt` variable is cleared, loaded with the value of d, or incremented or decremented based on the value of the control signals.

Register Inference also allows implementation of a latch. The latch is inferred from the incompletely specified IF Statement inside a process statement. Example 7.18 shows an implementation of a latch.

Example 7.18 Latch Implementation.

```
ENTITY latch IS
        PORT (enable, data: IN BIT;
                q: OUT BIT);
END latch;

ARCHITECTURE arch OF latch IS
BEGIN


latch: PROCESS (enable, data)
        BEGIN
```

```
            IF (enable = '1') THEN
                  q <= data;
            END IF;
      END PROCESS latch;
END arch;
```

If enable = '1', the value data is assigned to q. However, the IF Statement does not specify what happens if enable = '0'. In this case, the circuit maintains the previous state and creates a latch.

State Machines Synthesis

To describe a state machine, an Enumeration Type for states, a process statement for the state register, and the next-state logic can be used. Example 7.19 is the VHDL Design File that implements a 2-state state machine from the Figure 7.9.

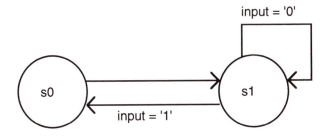

Figure 7.9 An example 2-state state machine

Example 7.19 2-State State Machine VHDL File.

```
ENTITY state_machine IS
        PORT (clk, reset, input: IN BIT;
                  output: OUT BIT);
END state_machine;

ARCHITECTURE arch OF state_machine IS

        TYPE STATE_TYP IS (s0, s1);
```

```
        SIGNAL state: STATE_TYP;

BEGIN

    PROCESS (clk, reset)

    BEGIN

            IF reset = '1' THEN
                    state <= s0;

            ELSIF (clk'EVENT AND clk = '1') THEN
                    CASE state IS

                            WHEN s0 =>
                                    state <= s1;

                            WHEN s1 =>
                                    IF input = '1' THEN
                                            state <= s0;

                                    ELSE
                                            state <= s1;
                                    END IF;

                    END CASE;

            END IF;

    END PROCESS;

    output <= '1' WHEN state = s1 ELSE '0';

END arch;
```

The process statement in Example 7.19 is sensitive to the clk and reset control signals. An IF statement inside the process statement is used to prioritize the clk and reset signals, giving reset the higher priority. The GDF equivalent of the state machine from the preceding example is shown in Figure 7.10.

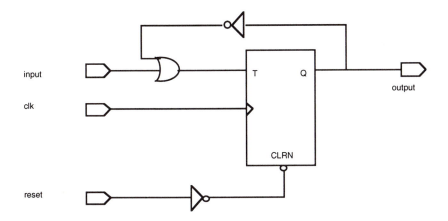

Figure 7.10 GDF equivalent of 2-state state machine

In general, the VHDL compiler assigns the value 0 to the first state, the value 1 to the second state, the value 2 to the third state, and so on. This state assignment can be overridden by manual state assignment using ENUM_ENCODING attribute which follows the associated Type declaration. Example 7.20 shows the manual state assignment.

Example 7.20 Manual State Assignment.

```
LIBRARY ieee;
USE ieee.std_logic_1164.all;

ENTITY state_machine IS
        PORT (up_down, clock: IN STD_LOGIC;
                lsb, msb: OUT STD_LOGIC);
END state_machine;
ARCHITECTURE enum_state_machine IS
        TYPE STATE_TYP IS (zero, one, two, three);
        ATTRIBUTE ENUM_ENCODING: STRING;
        ATTRIBUTE ENUM_ENCODING OF STATE_TYP: TYPE IS
                "11 01 10 00";
        SIGNAL present_state, next_state: STATE_TYP;
BEGIN
        PROCESS (present_state, up_down)
```

```
    BEGIN
        CASE present_state IS
            WHEN zero =>
                    IF up_down = '0' THEN
                            next_state <= one;
                            lsb <= '0';
                            msb <= '0';
                    ELSE
                            next_state <= three;
                            lsb <= '1';
                            msb <= '1';
                    END IF;
            WHEN one =>
                    IF up_down = '0' THEN
                            next_state <= two;
                            lsb <= '1';
                            msb <= '0';
                    ELSE
                            next_state <= zero;
                            lsb <= '0';
                            msb <= '0';
                    END IF;
            WHEN two =>
                    IF (up_down = '0') THEN
                            next_state <= three;
                            lsb <= '0';
                            msb <= '1';
                    ELSE
                            next_state <= one;
                            lsb <= '1';
                            msb <= '0';
                    END IF;
            WHEN three =>
                    IF (up_down = '0') THEN
                            next_state <= zero;
                            lsb <= '1';
                            msb <= '1';
                    ELSE
                            next_state <= two;
                            lsb <= '0';
                            msb <= '1';
                    END IF;
        END CASE;
```

```
        END PROCESS;

PROCESS
BEGIN
        WAIT UNTIL clock'EVENT AND clock = '1';
        present_state <= next_state;
END PROCESS;

END enum_state_machine;
```

The ENUM_ENCODING attribute must be a string literal that contains a series of state assignments. These state assignments are constant values that correspond to the state names in the Enumeration Type Declaration. The states in Example 7.20 above are encoded as shown in Table 7.2.

Table 7.2 State Encoding for State machine of Example 7.20.

zero	'11'
one	'01'
two	'10'
three	'00'

The ENUM_ENCODING attribute is Max+PLUS II specific and may not be available with other VHDL tools.

7.2.3 Hierarchical Projects

A VHDL Design File can be combined with the other VHDL Design Files and design files from other various tools (AHDL Design Files, GDF Design files, OrCAD Schematic Files) and some other vendor specific design files into a hierarchical project at any level of project hierarchy.

The Max+PLUS II design environment provides a number of primitives, bus macrofunctions, architecture-optimized macrofunctions, and application-specific macrofunctions. The designer can use Component Instantiation Statements to insert instances of macrofunctions and primitives. Register Inference can be used to implement registers.

Max+PLUS II Primitives

Max+PLUS II primitives are basic functional blocks used in circuit designs. Component Declarations for these primitives are provided in the maxplus2 package altera library in the maxplus2\max2vhdl\altera directory. Table 7.3 shows primitives that can be used in VHDL Design Files.

Table 7.3 Primitives supported by Max+PLUS II VHDL

Primitive Type	Primitive Name
Buffer	CARRY, CASCADE, EXP, GLOBAL, LCELL, SOFT, TRI
Flip-Flop	DFF, DFFE, JKFF, JKFFE, SRFF, SRFFE, TFF, TFFE
Latch	LATCH

Max+PLUS II Macrofunctions

Max+PLUS II macrofunctions are collections of high-level building blocks that can be used in logic designs. Macrofunctions are automatically installed in the \maxplus2\max2lib directory. Component Declarations for these macrofunctions are provided in the maxplus2 package in the altera library in the \maxplus2\max2vhdl\altera directory. The compiler analyses logic circuits and automatically removes all unused gates and D flip-flops. All input ports have default signal values, so the designer can simply leave unused inputs unconnected. From a functional point of view, all macrofunctions are the same regardless of the target architecture. However, implementations take advantages of the architecture of each device family, providing higher performance and more efficient implementation. Examples of Max+PLUS II macrofunctions supported by VHDL are shown in Table 7.4 and the rest can be found in the corresponding Altera literature. Macrofunction names usually have a prefix a_ due to the fact that VHDL does not support names that begin with digits.

Table 7.4 Max+PLUS II Macrofunctions supported by VHDL

Macrofunction Type	Macrofunction Name	Description of Operation
Adder	a_8fadd a_7480 a_74283	8-bit full adder Gated full adder 4-bit full adder with fast carry
Arithmetic Logic Unit	a_74181 a_74182	Arithmetic logic unit Look-ahead carry generator
Application specific	ntsc	NTSC video control signal generator
Buffer	a_74240 a_74241	Octal inverting 3-state buffer Octal 3-state buffer
Comparator	a_8mcomp a_7485 a_74688	8-bit magnitude comparator 4-bit magnitude comparator 8-bit identity comparator
Converter	a_74184	BCD-to-binary converter
Counter	gray4 a_7468 a_7493 a_74191 a_74669	Gray code counter Dual decade counter 4-bit binary counter 4-bit up/down counter with asynch. load Synchr. 4-bit up/down counter

Decoder	a_16dmux a_7446 a_74138	4-to-16 decoder BCD-to-7-segment decoder 3-to-8 decoder
EDAC	a_74630	16-bit parallel error detection &correction circuit
Encoder	a_74148 a_74348	8-to-3 encoder 8-to-3 priority encoder with 3-state outputs
Frequency divider	a_7456	Frequency divider
Latch	inpltch a_7475 a_74259 a_74845	Input latch 4-bit bistable latch 8-bit addressable latch with Clear 8-bit bus interface D latch with 3-state outputs
Multiplier	mult4 a_74261	4-bit parallel multiplier 2-bit parallel binary multiplier
Multiplexer	a_21mux a_74151 a_74157 a_74356	2-to-1 multiplexer 8-to-1 multiplexer Quad 2-to-1 multiplexer 8-to-1 data selector/multiplexer/register with 3-state outputs
Parity generator/checker	a_74180	9-bit odd/even parity generator/checker

Register	a_7470 a_7473 a_74171 a_74173 a_74396	AND-gated JK flip-flop with Preset and Clear Dual JK flip-flop with Clear Quad D flip-flops with Clear 4-bit D register Octal storage register
Shift register	barrelst a_7491 a_7495 a_74198 a_74674	8-bit barrel shifter Serial-in serial-out shift register 4-bit parallel-access shift register 8-bit bi-directional shift register 16-bit shift register
Storage register	a_7498	4-bit data selector/storage register
SSI Functions	inhb a_7400 a_7421 a_7432 a_74386	Inhibit gate NAND2 gate AND4 gate OR2 gate Quadruple XOR gate
True/Complement I/O Element	a_7487 a_74265	4-bit true/complement I/O element Quadruple complementary output elements

The Component Instantiation statement can be used to insert an instance of a Max+PLUS II primitive or macrofunction in circuit design. This statement also connects macrofunction ports to signals or interface ports of the associated entity/architecture pair. The ports of primitives and macrofunctions are defined with Component Declarations elsewhere in the file or in referenced packages. Consider Example 7.21.

Example 7.21 Port Definitions.

```
LIBRARY altera;
USE altera.maxplus2.ALL;

LIBRARY ieee;
USE ieee.std_logic_1164.ALL;

ENTITY example IS
        PORT (data, clock, clearn, presetn: IN STD_LOGIC;
                q_out:                          OUT STD_LOGIC;

                a, b, c, gn: IN STD_LOGIC;
                d:                       IN STD_LOGIC_VECTOR(7 DOWNTO
0);
                y, wn:       OUT STD_LOGIC);
END example;

ARCHITECTURE arch OF example IS

BEGIN

        dff1: DFF PORT MAP (d=>data, q=>q_out, clk=>clock,
                                clrn=>clearn, prn=>presetn);
        mux: a_74151b PORT MAP (c, b, a, d, gn, y, wn);
END arch;
```

Component Instantiation statements are used to create a DFF primitive and a 74151b macrofunction. The library altera is declared as the resource library. The USE clause specifies the maxplus2 package contained in the altera library. Figure 7.11 shows a GDF equivalent to the Component Instantiation Statements of the preceding example.

Besides using Max+PLUS II primitives and macrofunctions, a designer can implement the user defined macrofunctions with one of the following methods:

• Declare a package for each project-containing Component Declaration for all lower-level entities- in the top-level design file.

• Declare a component in the architecture in which it is instantiated.

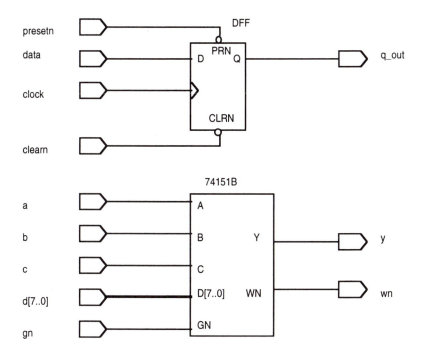

Figure 7.11 A GDF Equivalent of Component Instantiation Statement

Example 7.22 shows reg12.vhd, a 12-bit register that will be instantiated in a VHDL Design File at a higher level of design hierarchy.

Example 7.22 A 12-bit Register, reg12.vhd.

```
ENTITY reg12 IS
        PORT (d:        IN BIT_VECTOR (11 DOWNTO 0);
                clk: IN BIT;
                q:      OUT BIT_VECTOR (11 DOWNTO 0));
END reg12;

ARCHITECTURE arch OF reg12 IS
BEGIN
        PROCESS
        BEGIN
                WAIT UNTIL clk = '1';
                q <= d;
        END PROCESS;
END arch;
```

Figure 7.12 shows a GDF File equivalent to the preceding VHDL Example 7.22.

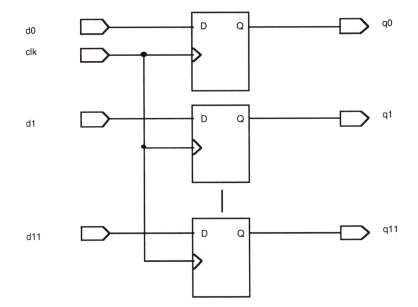

Figure 7.12 GDF Equivalent of reg12 Register

Example 7.23 declares the reg24_package, identifies it with a USE Clause, and uses the reg12 register as a component without requiring an additional Component Declaration.

Example 7.23 Declaration of the reg24_package.

```
PACKAGE reg24_package IS
      COMPONENT reg12
            PORT (d: IN BIT_VECTOR(11 DOWNTO 0);
                     clk: IN BIT;
                     q: OUT BIT_VECTOR(11 DOWNTO 0));
      END COMPONENT;
END reg24_package;

LIBRARY work;
USE work.reg24_package.ALL;
ENTITY reg24 IS
      PORT( d: IN BIT_VECTOR(23 DOWNTO 0);
                  clk: IN BIT;
                  q: OUT BIT_VECTOR(23 DOWNTO 0));
END reg24;

ARCHITECTURE arch OF reg24 IS
BEGIN
      reg12a: reg12 PORT MAP (d => d(11 DOWNTO 0),
                  clk => clk, q => q(11 DOWNTO 0));
      reg12b: reg12 PORT MAP (d => d(23 DOWNTO 12),
                  clk => clk, q => q(23 DOWNTO 12));
END arch;
```

From Example 7.23 we see that the user-defined macrofunction is instantiated with the ports specified in a Component Declaration. In contrast, Max+PLUS II macrofunctions are provided in the maxplus2 package in the altera library. The Architecture Body for reg24 contains two instances of reg12. A GDF example of the preceding VHDL file is shown in Figure 7.13.

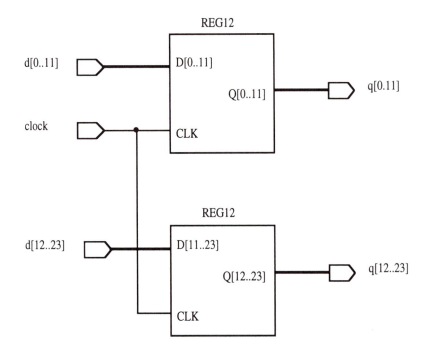

Figure 7.13 A GDF Equivalent of reg24

All VHDL libraries must be compiled. In Max+PLUS II, compilation is performed either with the Project Save and Compile command in any Max+PLUS II application or with the START button in the Max+PLUS II Compiler window.

7.3 Design Example in VHDL - Sequence Classifier and Recognizer

The purpose of this section is to present an example of a design using VHDL as a high level design tool. The design is a sequence classification and recognition circuit which receives a sequence of binary coded decimal numbers and counts those with the number of ones greater than or equal to the number of zeros in their codes, and the number of sequences with the number of ones less than number of zeros in their codes. The counting process stops when a specific sequence of input numbers is encountered. The system maintains two counters: the counter that stores the number of codes in which the number of ones has been greater than or equal to the number of zeros and the counter that stores the number of codes in which the number of ones has been less than the number of zeros. The counting, and classification of codes in the input sequence continues until a specific five digit sequence (password) is received, in which case the process of classification and counting stops.

The overall sequence classifier and recognizer is illustrated in Figure 7.14. The input sequence appears on both inputs of the classifier and recognizer. Classification activates one of two outputs that increment the counters. The recognizer is permanently analyzing the last five digits received in the sequence. When a specific sequence (given in advance) is recognized, the output of the recognizer is activated. This output stops counting on both counters.

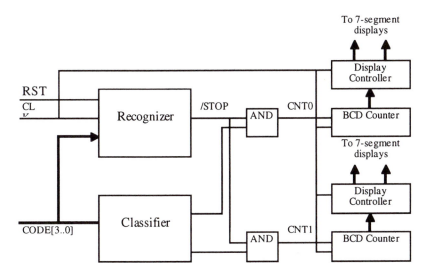

Figure 7.14 The block diagram of sequence classifier and recognizer circuit

The counters are of the BCD type providing three BCD-coded values on their outputs. These outputs are used to drive 7-segment displays, so that the current value of each counter is continuously displayed. In order to reduce the display control circuitry, three digits are multiplexed to the output of the display control circuitry also, a 7-segment LED enable signal determines which seven segment display output value is activated. Before displaying, values are converted into a 7-segment code.

7.3.1 Input Code Classifier

The input code classifier is a simple circuit that accepts on its input 4-bit code of a decimal digit represented by the BIT_VECTOR type input variable code. As a result of classification one of two output variables is activated:

- more_ones when the number of ones in the input code is grater than or equal to the number of zeros

- more_zeros when the number of ones is less than the number of zeros in the code

The VHDL description of this circuit (Example 7.24) consists of two processes. One process called counting counts the number of ones in an input code and exports that number in the form of the signal no_of_ones. Another process, called comparing, compares the number of ones with 2 and determines which output signal will be activated.

Example 7.24 Input Code Classifier VHDL File.

```
LIBRARY ieee;
USE ieee.std_logic_1164.ALL;

ENTITY class IS
        PORT
        (
                code                          :   IN   BIT_VECTOR(3
DOWNTO 0);
                more_ones                    : OUT STD_LOGIC;
                more_zeros                   : OUT STD_LOGIC
        );
END class;

ARCHITECTURE classifier OF class IS
```

```
        SIGNAL no_of_ones      : INTEGER RANGE 0 TO 4;

BEGIN
counting: PROCESS(code)
              VARIABLE n: INTEGER RANGE 0 TO 4;
        BEGIN
        n:=0;
        FOR i IN 0 TO 3 LOOP
              IF code(i)='1' THEN
                     n:=n+1;
              END IF;
        END LOOP;
        no_of_ones <= n;
        END PROCESS counting;

comparing: PROCESS(no_of_ones)
        BEGIN
        IF no_of_ones >= 2 THEN
              more_ones <= '1';
              more_zeros <= '0';
        ELSE
              more_ones <= '0';
              more_zeros <= '1';
        END IF;
        END PROCESS comparing;

END classifier;
```

The Max+PLUS II Compiler is able to synthesize the circuit from this behavioral description.

7.3.2 Sequence recognizer

The input sequence recognizer is practically the same as the electronic lock example of Section 5.1. In this case we present a solution for the simpler version of the recognizer. If the correct digit appears on its input, even repeated several times, the recognizer advances through the states until it comes to the final state with five recognized digits. In this case it activates the stop output which is used to stop counting on both BCD counters. Otherwise, when an incorrect digit is encountered, the recognizer returns to its initial state, start, in which it activates the restart signal indicating that a new recognition cycle has started.

The VHDL design of the input sequence recognizer is shown as Example 7.25. The recognizer is designed as a state machine that is easily synthesized by the Max+PLUS II Compiler.

Example 7.25 Input Sequence Recognizer.


```
LIBRARY ieee;
USE ieee.std_logic_1164.ALL;

ENTITY recognizer IS
PORT
(
        code                        : IN BIT_VECTOR(3 DOWNTO 0);
        clk, reset                  : IN STD_LOGIC;
        restart, stop         : OUT STD_LOGIC
);
END recognizer;


ARCHITECTURE fivedigits OF recognizer IS
        TYPE rec_state IS (start, one, two, three, four, five);
        SIGNAL state         : rec_state;
        CONSTANT d1: BIT_VECTOR :="1001";
        CONSTANT d2: BIT_VECTOR :="1000";
        CONSTANT d3: BIT_VECTOR :="0111";
        CONSTANT d4: BIT_VECTOR :="0110";
        CONSTANT d5: BIT_VECTOR :="0101";

BEGIN

PROCESS(clk)

        BEGIN

        IF (clk'EVENT AND clk='1') THEN
                IF reset = '1' THEN
                state <= start;
                ELSE
```

```
CASE state IS
      WHEN start =>
              IF (code= d1) THEN
                      state <= one;
              ELSE
                      state <= start;
              END IF;
                      restart <= '1';
                      stop <= '0';
      WHEN one =>
              IF (code=d2) THEN
                      state <= two;
              ELSIF (code=d1) THEN
                      state <= one;
              ELSE
                      state <= start;

              END IF;
              restart <='0';
              stop <= '0';

      WHEN two =>
              IF (code=d3) THEN
                      state <= three;
              ELSIF (code=d2) THEN
                      state <= two;
              ELSE
                      state <= start;
              END IF;
                      restart <= '0';
                      stop <= '0';

      WHEN three =>
              IF (code=d4) THEN
                      state <= four;
              ELSIF (code=d3) THEN
                      state <= three;
              ELSE
                      state <= start;
              END IF;
                      restart <= '0';
                      stop <= '0';
```

```
                        WHEN four =>
                            IF (code=d5) THEN
                                    state <= five;
                            ELSIF (code=d4) THEN
                                    state<= four;
                            ELSE
                                    state <= start;
                            END IF;
                                    restart <= '0';
                                    stop <= '0';

                        WHEN five =>
                                    state <= five;
                                    restart <= '0';
                                    stop <= '1';

                        END CASE;

        END IF;
        END IF;
END PROCESS;

END fivedigits;
```

The states of the state machine are declared as enumeration type rec_state and the current state, state, is of that type. The sequence which has to be recognized is declared in the architecture declaration part by constant declarations which can easily be changed to any desired sequence.

7.3.3 BCD Counter

The BCD counter presented in Example 7.26 is a three digit counter that consists of three simple modulo 9 counters connected serially. The individual counters are presented by the processes bcd0, bcd1, and bcd2, that communicate by way of internal signals cout and cin which are used to enable the counting process. The ena input is used to enable the counting process of the least significant digit counter. At the same time, it is used to enable the entire BCD counter. Each individual counter has to recognize 8 and 9 input changes to prepare itself and the next stage for the proper change.

Example 7.26 BCD Counter.

```
LIBRARY ieee;
USE ieee.std_logic_1164.ALL;

ENTITY bcdcount IS
PORT(
        clk, reset, ena                 : IN STD_LOGIC;
        dout0,dout1,dout2               : OUT INTEGER RANGE 0 TO 9;
        cout                            : OUT STD_LOGIC
);
END bcdcount;

ARCHITECTURE bcdcounter OF bcdcount IS
        SIGNAL cout0,cout1,cin1,cin21, cin20 : STD_LOGIC;
BEGIN

bcd0: PROCESS(clk)
                VARIABLE n: INTEGER RANGE 0 TO 9;
        BEGIN

                IF (clk'EVENT AND clk='1') THEN
                        IF reset = '1' THEN
                                n:=0;
                        ELSE
                                IF ena = '1' AND n<9 THEN
                                        IF n=8 THEN
                                        n := n+1;
                                        cout0 <= '1';
                                        ELSE
                                        n:=n+1;
                                        cout0 <= '0';
                                        END IF;
                                ELSIF ena = '1' AND n=9 THEN
                                        n := 0;
                                        cout0 <= '0';
                                END IF;
                        END IF;
                END IF;
                        dout0 <= n;
        END PROCESS bcd0;
```

```
bcd1: PROCESS(clk)
            VARIABLE n: INTEGER RANGE 0 TO 9;
        BEGIN
        cin1 <= cout0;
            IF (clk'EVENT AND clk='1') THEN
                    IF reset = '1' THEN
                            n := 0;
                    ELSE
                            IF cin1 = '1' AND n<9 THEN
                                    IF n=8 THEN
                                    n := n+1;
                                    cout1 <= '1';
                                    ELSE
                                    n:=n+1;
                                    cout1 <= '0';
                                    END IF;
                            ELSIF cin1 = '1' AND n=9 THEN
                                    n := 0;
                                    cout1 <= '0';
                            END IF;
                    END IF;
            END IF;
                    dout1 <= n;
        END PROCESS bcd1;


bcd2: PROCESS(clk)
            VARIABLE n: INTEGER RANGE 0 TO 9;
        BEGIN
        cin21 <= cout1;
        cin20 <= cout0;
            IF (clk'EVENT AND clk='1') THEN
                    IF reset = '1' THEN
                            n := 0;
                    ELSE
                            IF cin21 = '1' AND cin20 = '1' AND
n<9 THEN
                                    IF n=8 THEN
                                    n := n+1;
                                    cout <= '1';
                                    ELSE
                                    n:=n+1;
                                    cout <= '0';
                                    END IF;
```

```
                              ELSIF cin21= '1' AND cin20 = '1' AND
n=9 THEN
                                    n := 0;
                                    cout <= '0';
                              END IF;

                        END IF;
                  END IF;
                        dout2 <= n;
            END PROCESS bcd2;

END bcdcounter;
```

7.3.4 Display Controller

The display controller carries out the same operations as the controller shown in Section 5.2. It receives three binary coded decimal digits on its inputs, passing one digit at time to the output while activating the signal that determines which 7-segment display the digit will be forwarded. It also performs conversion of binary into 7-segment code. The VHDL description of the display control circuitry is shown in Example 7.27. It consists of three processes. The first process, count, implements a modulo 2 counter that selects, in turn, three input digits to be displayed. In the second process, the counter's output is used to select which digit is passed through the multiplexer, represented by the mux process, and at the same time selects which 7-segment display the digit will be displayed. The third process, called converter, performs code conversion from binary to 7-segment code.

Example 7.27 Display Control Circuitry.

```
LIBRARY ieee;
USE ieee.std_logic_1164.ALL;

ENTITY displcont IS
PORT
(
      dig0, dig1, dig2              : IN INTEGER RANGE 0 TO 9;
      clk                              :              I N
STD_LOGIC;
```

```
        sevseg                           :  OUT  BIT_VECTOR(6
DOWNTO 0);
        ledsel0, ledsel1, ledsel2    : OUT STD_LOGIC
);
END displcont;

ARCHITECTURE displ_beh OF displcont IS
        SIGNAL q, muxsel      : INTEGER RANGE 0 TO 2;
        SIGNAL bcd            : INTEGER RANGE 0 TO 9;
BEGIN

count: PROCESS(clk)
        VARIABLE n:    INTEGER RANGE 0 TO 2;
BEGIN
        IF(clk'EVENT AND clk='1') THEN
                IF n < 2 THEN
                        n:= n+1;
                ELSE
                        n:=0;
                END IF;
                IF n=0 THEN
                        ledsel0 <= '1';
                        ledsel1 <= '0';
                        ledsel2 <= '0';
                        q <= n;
                ELSIF n=1 THEN
                        ledsel1<='1';
                        ledsel0 <= '0';
                        ledsel2 <= '0';
                        q <=n;
                ELSIF n=2 THEN
                        ledsel2 <= '1';
                        ledsel0 <= '0';
                        ledsel1 <= '0';
                        q <=n;
                ELSE
                        ledsel0 <= '0';
                        ledsel1 <= '0';
                        ledsel2 <= '0';
                END IF;
        END IF;
END PROCESS count;
```

```
mux: PROCESS(dig0, dig1, dig2, muxsel)
BEGIN
muxsel <=q;
      CASE muxsel IS
              WHEN 0 =>
                      bcd <= dig0;
              WHEN 1 =>
                      bcd <= dig1;
              WHEN 2 =>
                      bcd <= dig2;
              WHEN OTHERS =>
                      bcd <= 0;
      END CASE;
      END PROCESS mux;


converter: PROCESS(bcd)
BEGIN
      CASE bcd IS
              WHEN 0 => sevseg <= "1111110";
              WHEN 1 => sevseg <= "1100000";
              WHEN 2 => sevseg <= "1011011";
              WHEN 3 => sevseg <= "1110011";
              WHEN 4 => sevseg <= "1100101";
              WHEN 5 => sevseg <= "0110111";
              WHEN 6 => sevseg <= "0111111";
              WHEN 7 => sevseg <= "1100010";
              WHEN 8 => sevseg <= "1111111";
              WHEN 9 => sevseg <= "1110111";
              WHEN others => sevseg <= "1111110";
      END CASE;
END PROCESS converter;

END displ_beh;
```

7.3.5 Circuit Integration

The sequence classifier and recognizer (Example 7.28) is integrated using existing components and structural modeling. All components are declared in the architecture declaration part and then instantiated the required number of times. The interconnections of components are achieved using internal signals

declared in the architecture declaration part of the design. As the result of its
operation the overall circuit provides two sets of 7-segment codes directed to 7-
segment displays, together with the enable signals which select a 7-segment
display which the resulting code is directed.

Example 7.28 Sequence Classifier and Recognizer.


```
LIBRARY ieee;
USE ieee.std_logic_1164.ALL;

ENTITY iscr IS
PORT
(
        code                    : IN BIT_VECTOR(3 DOWNTO 0);
        clk, rst                : IN STD_LOGIC;
        sevsega, sevsegb        : OUT BIT_VECTOR(6 DOWNTO 0);
        restart                 : OUT STD_LOGIC;
        endr                    : OUT STD_LOGIC;
        leda0, leda1, leda2     : OUT STD_LOGIC;
        ledb0, ledb1, ledb2     : OUT STD_LOGIC;
        overfl0, overfl1        : OUT STD_LOGIC
);
END iscr;

ARCHITECTURE structural OF iscr IS
        SIGNAL cnt0, cnt1               : STD_LOGIC;
        SIGNAL clas0, clas1             : STD_LOGIC;
        SIGNAL succ                     : STD_LOGIC;
        SIGNAL d0out0, d0out1, d0out2 : INTEGER RANGE 0 TO 9;
        SIGNAL d1out0, d1out1, d1out2 : INTEGER RANGE 0 TO 9;

        COMPONENT recogniser
        PORT
        (
                code                    : IN BIT_VECTOR(3 DOWNTO 0);
                clk, reset              : IN STD_LOGIC;
                restart, success        : OUT STD_LOGIC
        );
        END COMPONENT;

        COMPONENT bcdcount
```

```
        PORT
        (
                clk, reset, ena         : IN STD_LOGIC;
                dout0,dout1,dout2       : OUT INTEGER RANGE 0 TO 9;
                cout                    : OUT STD_LOGIC
        );
        END COMPONENT;

        COMPONENT displcont
        PORT
        (
                dig0, dig1, dig2        : IN INTEGER RANGE 0 TO 9;
                clk                     : IN STD_LOGIC;
                sevseg          : OUT BIT_VECTOR(6 DOWNTO 0);
                ledsel0, ledsel1, ledsel2       : OUT STD_LOGIC
        );
        END COMPONENT;

        COMPONENT class
        PORT
        (
                input_code              : IN BIT_VECTOR(3 DOWNTO 0);
                more_ones               : OUT STD_LOGIC;
                more_zeros              : OUT STD_LOGIC
        );
        END COMPONENT;

BEGIN
        cnt0 <= clas0 AND (NOT succ);
        cnt1 <= clas1 AND (NOT succ);
        endr <= succ;

        recogn:recogniser PORT MAP (code, clk, rst, restart,
succ);
        classif: class PORT MAP (code, clas1, clas0);
        bcdcnt0: bcdcount PORT MAP (clk, rst, cnt0, d0out0,
d0out1,
                        d0out2, overfl0);
        bcdcnt1: bcdcount PORT MAP (clk, rst, cnt1, d1out0,
d1out1,
                        d1out2, overfl1);
        disp0: displcont PORT MAP (d0out0, d0out1, d0out2, clk,
                        sevsega, leda0, leda1, leda2);
```

```
disp1: displcont PORT MAP (dlout0, dlout1, dlout2, clk,
                    sevsegb, ledb0, ledb1, ledb2);
```

END structural;

This design was successfully compiled into one FLEX 8450 device and also compiled into one MAX 7096 device. With a small change of the logic of the display controller, the design also fits into the smallest FLEX device, FLEX 8282. This change uses only one binary to 7-segment converter as illustrated in Figure 7.15. The digits from the BCD counters are brought to the common multiplexer where only one is selected to be displayed and the corresponding LED selection signal is activated. The VHDL of the modified display controller is given below as Example 7.29.

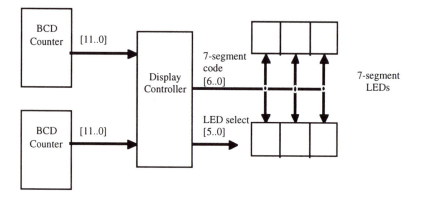

Figure 7.15 The modified display controller

Example 7.29 Modified Display Controller.

```
LIBRARY ieee;
USE ieee.std_logic_1164.ALL;

ENTITY displcon1 IS
```

```
PORT
(
        adig0, adig1, adig2    : IN INTEGER RANGE 0 TO 9;
        bdig0, bdig1, bdig2    : IN INTEGER RANGE 0 TO 9;
        clk                    : IN STD_LOGIC;
        sevseg                 : OUT BIT_VECTOR(6 DOWNTO 0);
        aledsel0, aledsel1, aledsel2 : OUT STD_LOGIC;
        bledsel0, bledsel1, bledsel2 : OUT STD_LOGIC
);
END displcon1;

ARCHITECTURE displ_beh OF displcon1 IS
        SIGNAL q, muxsel       : INTEGER RANGE 0 TO 5;
        SIGNAL bcd             : INTEGER RANGE 0 TO 9;

BEGIN
count: PROCESS(clk)
        VARIABLE n:    INTEGER RANGE 0 TO 5;
BEGIN
        IF(clk'EVENT AND clk='1') THEN
                IF n < 5 THEN
                        n:= n+1;
                ELSE
                        n:=0;
                END IF;
                IF n=0 THEN
                        aledsel0 <= '1';
                        aledsel1 <= '0';
                        aledsel2 <= '0';
                        q <= n;
                ELSIF n=1 THEN
                        aledsel0 <= '0';
                        aledsel1 <= '1';
                        aledsel2 <= '0';
                        q <=n;
                ELSIF n=2 THEN
                        aledsel0 <= '0';
                        aledsel1 <= '0';
                        aledsel2 <= '0';
                        q <=n;
                ELSIF n=3 THEN
                        bledsel0 <= '1';
                        bledsel1 <= '0';
```

```
                              bledsel2 <= '0';
                              q <=n;
                    ELSIF n=4 THEN
                              bledsel0 <= '0';
                              bledsel1 <= '1';
                              bledsel2 <= '0';
                              q <=n;
                    ELSE
                              bledsel0 <= '0';
                              bledsel1 <= '0';
                              bledsel2 <= '1';
                    END IF;
            END IF;
    END PROCESS count;

    mux: PROCESS(adig0, adig1, adig2, bdig0, bdig1, bdig2, muxsel)
    BEGIN
    muxsel <=q;
            CASE muxsel IS
                    WHEN 0 =>
                              bcd <= adig0;
                    WHEN 1 =>
                              bcd <= adig1;
                    WHEN 2 =>
                              bcd <= adig2;
                    WHEN 3 =>
                              bcd <= bdig0;
                    WHEN 4 =>
                              bcd <= bdig1;
                    WHEN 5 =>
                              bcd <= bdig2;
            END CASE;
            END PROCESS mux;
    converter: PROCESS(bcd)
    BEGIN
            CASE bcd IS
                    WHEN 0 => sevseg <= "1111110";
                    WHEN 1 => sevseg <= "1100000";
                    WHEN 2 => sevseg <= "1011011";
                    WHEN 3 => sevseg <= "1110011";
                    WHEN 4 => sevseg <= "1100101";
                    WHEN 5 => sevseg <= "0110111";
                    WHEN 6 => sevseg <= "0111111";
```

```
          WHEN 7 => sevseg <= "1100010";
          WHEN 8 => sevseg <= "1111111";
          WHEN 9 => sevseg <= "1110111";
          WHEN others => sevseg <= "1111110";
     END CASE;
     END PROCESS converter;

END displ_beh;
```

8 RAPID PROTOTYPING OF COMPUTER SYSTEMS USING FPLDS- A CASE STUDY

Rapid prototyping systems composed of programmable components show great potential for full implementation of microelectronics designs. Prototyping systems based on field programmable devices present many technical challenges affecting system utilization and performance.

This chapter addresses two key issues to assess and exploit today's rapid-prototyping methodologies. The first issue is the development of architectural organizations to integrate field-programmable logic with an embedded microprocessor (Intel 386 EX) as well as system integration issues. The second is the design of prototyping systems as Custom Computing Engines.

Prototyping systems can potentially be extended to general custom computing machines in which the architecture of the computer evolves over time, changing to fit the needs of each application it executes. In particular, we will focus on implementing Private Eye display control, PSRAM control, and some secondary logic (PCMCIA control) using FPLD.

8.1 System Overview

The VuMan family of wearable computers, developed at the Carnegie Mellon University Engineering Design Research Center, will be used as the platform. The latest product in the line, VuMan 3, mixes off-the-shelf hardware components with software developed in-house to form an embedded system used by the US Marines for military vehicle maintenance. VuMan 3 specializes in maintenance applications for environments requiring rugged, portable electronic tools.

The components necessary to accommodate such needs include processing core, memory, BIOS/Bootcode ROM, display adapter, direct-memory-access (DMA) controller, serial ports, real-time clock, power control, input system, and peripheral controller. A functional diagram depicting the VuMan architecture is shown in Figure 8.1.

An Intel i386EX embedded microprocessor acts as the system core. This 3.3 volt version of the 386 carries several critical system components on-chip, such as DMA controller, interrupt controller, timers, serial ports, chip-select unit, and refresh unit. This makes it an ideal solution for the VuMan 3 embedded system since exploiting the on-chip services helps reduce chip count and board area. Additionally, the processor provides several signals allowing seamless connections to memories and I/O devices. This feature further reduces the chip-count by eliminating the need for CPU-to-memory interface logic.

The memory subsystem attached to this processor consists of two components. Two Hitachi 3.3 volt 512K P-SRAMs (pseudo-static RAMs) provide a 16-bit path to 1MB of main memory (2 chips, 8 bits each = 16 bits). One chip stores all data on even addresses and the other maintains all odd address bytes. Likewise, two Hitachi 3.3 volt 128K SRAMs offer a 16-bit interface to 256K of RAM. Moreover, these memories obtain power from a battery, not the system power source, so they can be used to store vital data. They require no special interface and can attach gluelessly to the CPU core. Also, these memories use static cells eliminating the need for refresh logic.

The 512K chips, on the other hand, require periodic refreshing since they are not purely static RAMs. The 386EX's refresh control until assists in performing this function. Also, these memories require a pre-charge between accesses, which eliminates the possibility of using the direct CPU-to-memory interface offered by the i386EX. Therefore, this subsystem requires a control system to act as an interface between the system bus and the memories.

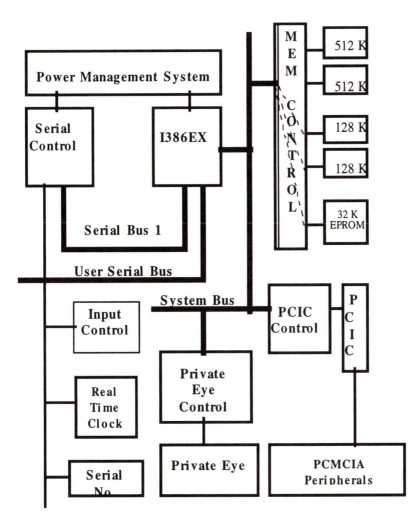

Figure 8.1 VuMan Functional Diagram

An additional non-volatile memory element, the 3.3 volt 32K EPROM, contains the code necessary to boot the system and initialize the hardware. The processor begins executing from the EPROM on power-up and the code stored therein must configure the system as desired. The bootcode sets up the system memory map as shown in Figure 8.2, performs testing of critical components, and then transfers control to the user application. The i386EX bus interface accommodates the EPROM seamlessly; hence, the ROM attaches directly to the system bus.

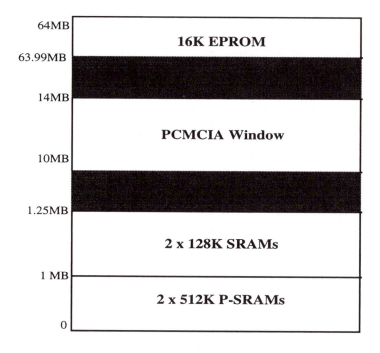

Figure 8.2 VuMan 3 Memory Map

A Reflection Technology Private Eye provides the system with a 720x280-pixel display. This device allows the system to serially deliver pixel data to be drawn. The Private Eye, however, uses 5V signals so care must be taken when interfacing this device to the 3.3V system bus. The serial protocol used by the Private Eye requires the development of an interface to manage the communications between the processor and the display adapter.

The input system consists of a dial and buttons. When the user presses any of the buttons or rotates the dial, a code is sent to the processor by way of the i386EX's synchronous serial port, and the processor reacts accordingly. Likewise, when the CPU wishes to read the real-time clock reading or the silicon serial number, these values flow to the 386EX through the serial port. A PIC microcontroller manages the serial protocol between the processing core and these devices. This PIC also manages the power supplies and informs the processor when the power reaches dangerously low levels.

Lastly, the system uses the Intel 8256 PCIC (PCMCIA controller) chip to manage the two PCMCIA slots in the system. These slots allow for connections of memory cards or peripheral cards. The PCIC must be programmed to map the PCMCIA cards into a certain memory region. Then, when the PCIC detects accesses to this region, it forwards the requests to the appropriate slot. Since the PCIC uses 5-volt signals, an interface must exist to manage the communications between the 5-volt PCIC and the 3.3-volt system bus. Also, as the PCIC expects ISA-compatible signals, the interface must convert the 386EX bus signals into semantically identical ISA signals.

8.2 Memory Interface Logic

Though the 386EX was designed to allow effortless connection of memory devices to the system bus, this feature cannot be exploited in the VuMan3 design due to the P-SRAMs used. These RAMs require a very strict protocol when performing reads or writes (see Hitachi HM65V8512 datasheet for details): (a) the chips must be turned off (chip-enable deasserted) for 80 ns between consecutive accesses, (b) during reads, the chip-enable must be on for 15 ns before output-enable can be asserted to deliver the data, and (c) during writes, the write-enable signal must be pulsed for 35 ns while the chip-enable is on. Also, the chip-enable must remain active for 150 ns during each read/write cycle. Additionally, since these chips lack purely static cells, they require periodic refresh. A refresh cycle consists of pulsing the output-enable (OE) signal of the memory while keeping the chip-enable (CE) off. The following are timing requirements for the refresh cycle: a) CE must be off for 80 ns before OE can be asserted, b) OE must be off for at least 40 ns before it gets pulsed, and c) OE must be active for 80 ns during the pulse. To accommodate these requirements, the memory controller of Figure 8.3 was designed.

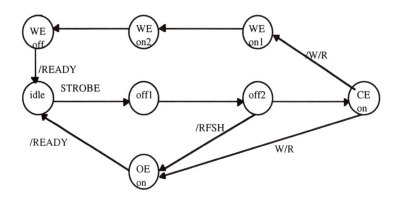

Figure 8.3 P-SRAM Memory Controller

The P-SRAM controller has seven input pins and five output pins defined in Table 8.1 ("/" denotes an active low signal).

Table 8.1 P-SRAM Controller Inputs and Outputs.

INPUTS:	
/ADS (ADdress Stable)	asserted by processor when a new bus cycle begins
/RFSH	asserted by processor when the cycle is a PSRAM refresh cycle
/CS1 (Chip-Select 1)	asserted by processor when it wishes to communicate with the PSRAMs
/BLE (Byte Low Enable)	asserted by processor if it intends to read/write

	the low-byte (data bits 0-7) of the addressed word
/BHE (Byte High Enable)	asserted by processor if it intends to read/write the high-byte (data bits 8-15) of the addressed word
W/R	high during write cycle, low during read cycle
/READY	asserted by 386EX's chip select unit when the bus cycle terminates (after X wait states have expired, where X is programmed by the bootcode)
STROBE	this signal is formed from the input signals as /ADS*[/RFSH+/CS1*(/BLE+?BHE)]
OUTPUTS:	
/PLB	(PSRAM_CE_LOWBYTE)turns on in state CEon if /BLE is asserted and remains on until state idle
/PHB	(PSRAM_CE_HIGHBYTE) turns on in state CEon if /BHE is asserted and remains on until state idle
/PWE	PSRAM_WE is on during states WEon1 and WEon2
/POE	PSRAM_OE is on during state Oeon
/PCE	PSRAM_CE turns on when off2 state finishes and remains on until state idle.

This state machine maintains synchronization with the system bus through the 32MHz processor clock (CLK2). Hence, a transition from one state to another occurs every $1/32MHz = 31.25ns$.

Initially, the state machine is idle and the memory chip-select signal remains off. When the processor issues a bus cycle, it first sets the appropriate chip-select and address lines and asserts the /ADS signal . The 512K P-SRAMs tie to the chip-select signal /CS1. Hence, if the i386EX turns on /CS1, it intends to communicate with the 512K RAMs. When the processor needs the low byte of the bus, it asserts the /BLE line and when it needs the high byte, it turns on /BHE. Similarly, for a 16-bit access both /BLE and /BHE are asserted. Hence, when /ADS asserts, if either /BLE or /BHE are asserted and /CS1 is on, the current bus cycle involves the P-SRAM memories; hence, the state machine activates at this point.

When this bus cycle begins, the memory controller transitions to the OFF1 state. Since the chip-select is off during the idle state, the memory is guaranteed to have been off for at least 31.25 ns by the time the state machine enters the OFF1 state. At this stage, the chip-select is kept off and a transition is made to the OFF2 state with the coming of the next clock. This increased the guaranteed memory deselected time to 31.25 ns + 31.25 ns = 62.50 ns. In the OFF2 state, the state machine transitions based on the current bus cycle type: read, write, or refresh.

Recall that a refresh cycle requires that the OE be pulsed while the CE is off. Therefore, if the current cycle is a refresh cycle, the state machine transitions to the OEon state, in which the OE signal turns on. Therefore, by the time the machine enters the OEon state, CE and OE have been off for 93.75ns, which satisfies the refresh timings. A transition back to the idle state happens as soon as the READY signal is asserted, signifying the end of the bus cycle. To meet the timing requirement, the OE pulse must last 80 ns, so the state machine needs to remain in the OEon state for at least 80 ns. Hence, the machine can return to the idle state after 173.75 ns have elapsed from the beginning of the bus cycle. Normally, bus accesses require 2 bus clock periods. During the first bus period (known as T1 [refer to the 386EX manual, ch.7: Bus Interface Unit]), the address and status signals are set by the processor. During the second period (known as T2) the device responds. If a peripheral needs more bus periods, it requests a wait states, each of which lasts for one bus period. The 386EX bus uses a 16 MHz clock, yielding a 1/16 MHz (62.50 ns) bus period. Hence, a normal cycle requires 2 bus periods (125 ns). Since the refresh requires 173.75 ns, it needs an additional 48.75 ns, or 1 wait state.

Read/write cycles proceed similarly. If the current bus cycle is not a refresh cycle, the machine transitions to state CEon. By the time the machine arrives at this state, the memory has been de-selected for 31.25 + 31.25 + 31.25 = 93.75 ns, which meets the pre-charging requirement of 80 ns. When the state machine enters this state, it turns on the chip-select unit for the appropriate memory, as determined by the /BLE and /BHE signals: /PLB is asserted if /BLE is on, /PHB is asserted If /BHE is on, and both are asserted if both /BLE and /BHE are on. Next, the state machine determines whether the cycle is a read or a write and it transitions accordingly.

During a read cycle, the machine enters the OEon state. In this state, the /POE is asserted and remains on until the cycle terminates, indicated by the processor asserting /READY. Hence, the CE is on for 31.25 ns before the OE turns on, satisfying the timing requirements. Additionally, CE must be on for 150ns to meet the access time requirement, so the state machine cannot return to the idle state until (when CE goes on) 243.75 ns have elapsed from the beginning of the cycle. Hence, for read/write, the memory needs 118.75 ns, or 2 wait states. The read bus cycle is depicted in the timing diagram of Figure 8.4.

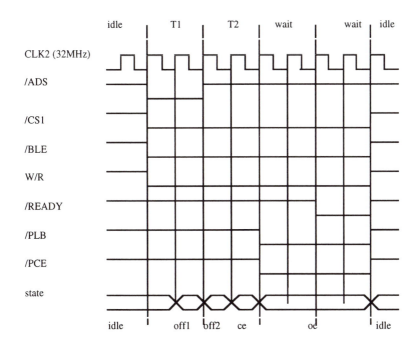

Figure 8.4 Read Cycle

Similarly, during a write cycle, the state machine proceeds from the CEon state to the WEon1 state. In this state, the /PWE signal is asserted, which starts the write to the memory. The machine transitions to the WEon2 on the next CLK2 edge and keeps /PWE active. Then, a transition to the WEoff state is made, and the /PWE is turned off. Hence, the /PWE is on for 2 (62.5 ns), meeting the timing requirement. The state machine remains in the Weoff state until the bus cycle is over, which sends the machine back to the idle state, where it turns off the memory chip-enables and awaits the next access to the memories. The write bus cycle is shown in Figure 8.5.

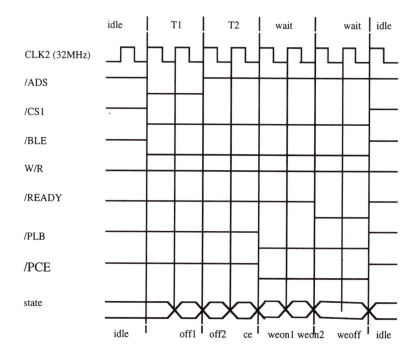

Figure 8.5 Write Cycle

Finally, the refresh bus cycle is shown in Figure 8.6.

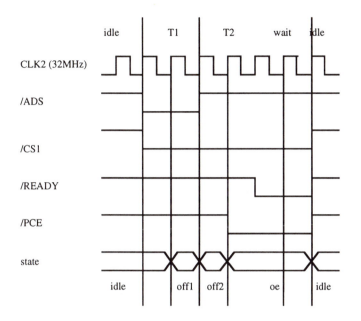

Figure 8.6 Refresh Cycle

This procedure allows for correct timing when accessing the P-SRAMs. The
state machine above interfaces the 386EX bus to the P-SRAM memories. It is
described using AHDL in Example 8.1.

Example 8.1 PSRAM Controller.

```
SUBDESIGN PSRAM_Controller
(
        clk, reset                              : INPUT;
        /ADS,/RFSH,/CS1,/BLE,/BHE,/W/R    : INPUT;
        /PCS,/PLB,/PHB                          :OUTPUT;
        /PWE,/POE                               :OUTPUT;
)
VARIABLE
```

```
        strobe,BE                              :NODE;
        ss        :M A C H I N E      W I T H      S T A T E S (i d d,
offl,off2,ce,oe,wel,we2,weoff);

BEGIN
        ss.clk=clk;
        ss.reset=reset;

        /PLB=!(((!/BLE)&BE)&/RFSH);
        /PHB=!(((!/BHE)&BE)&/RFSH);
        strobe =(!/ADS)&((!/RFSH)#((!/CS1)&((!/BLE)#(!/BHE))));


        TABLE

        ss, strobe, /RFSH,/W/R,/READY => ss,BE,/PWE,/POE,/PCS;

        idd,0,x,x,x => idd,0,1,1,1;
        idd,1,x,x,x => off1,0,1,1,1;
        off1,x,x,x,x =>off2,0,1,1,1;
        off2,x,0,x,x => oe,0,1,1,1;
        off2,x,1,x,x => ce,0,1,1,1;
        ce,x,x,1,x => wel,1,1,1,0;
        wel,x,x,1,x => we2,1,0,1,0;
        we2,x,x,1,x => weoff,1,0,1,0;
        weoff,x,x,x,1 => weoff,1,1,1,0;
        weoff,x,x,x,0 => idd,1,1,1,0;
        ce,x,x,0,x => oe,1,1,1,0;
        oe,x,x,x,0 => idd,1,1,0,0;
        oe,x,x,x,1 => oe,1,1,0,0;

        END TABLE;
END;
```

8.3 Private Eye Controller

The Private Eye (PE) Controller contains an 8-bit shift register used to receive data in parallel from the microprocessor that is to be displayed and subsequently deliver that information in the serial form to the PE display adapter. The structure of PE Controller is shown in Figure 8.7. Besides the shift register, the controller contains a frequency divider and a counter. The frequency divider divides the system frequency of 32MHz by four to provide the frequency required by the PE display. The counter counts a number of bits transmitted from the Controller and stops shifting process when all bits are delivered.

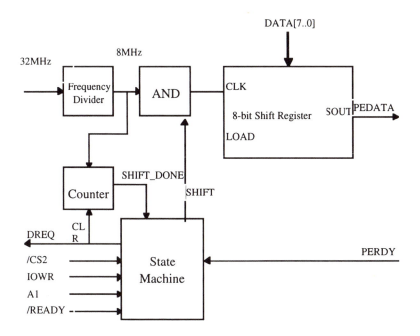

Figure 8.7 PE Controller Structure

Another state machine, described below, interfaces the system bus to the PE display adapter. It provides the PE Controller to be in one of four possible states:

- Idle (starting state from which it can be transferred into receiving state)

- Recv (receiving state in which it receives the next byte from the microprocessor to display it)

- Load (into which it comes upon receiving byte and loading it into shift register, and)

- Shift (the state in which it delivers byte to be displayed to the PE Controller bit by bit)

Its state transition diagram is depicted in Figure 8.8.

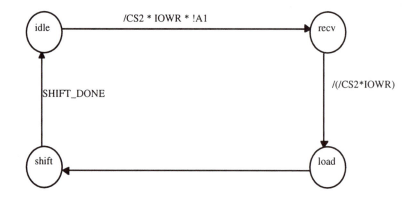

Figure 8.8 PE Controller State Transition Diagram

More detailed description of operation is given below. Signals used by the PE Controller are shown in Table 8.2.

Table 8.2 PE Controller Signals.

/CS2 (Chip-Select 2)	asserted by processor when it wishes to communicate with the PE control PLD
A1	bit 1 of the 26-bit address placed on the bus by the i386EX
IOWR	signifies that the current cycle is an I/O write cycle
SHIFT	asserted by the PE Controller when it enters the SHIFT state, indicating the PE Serial Data Shifter should begin delivering the data to the display adapter
SHIFT_DONE	asserted by PE Serial Data Shifter state machine indicating the shift is done
PERDY	asserted by PE when it is ready to accept data

The PE controller provides a mechanism by which the CPU can send pixel data to the display adapter. The PE (PE) display has 4 relevant signals: PEBOS, PERDY, PEDATA, and PECLK. The PEBOS (Beginning of Scan) tells the PE that new screen data is about to be delivered. The PERDY signal tells the controller when the PE is ready to receive data; when the PE refreshes the screen from its internal bitmap memory it deasserts PERDY and cannot accept data. The PEDATA and PECLK provide a data path from a host to the PE. The host places pixel data (1 bit = 1 pixel) on the PEDATA line and delivers it to the PE by clocking the PECLK. Since the screen consists of 720x280 pixels, the host must clock in 201,600 bits per screen. Also, the PE can accept data at a maximum rate of 8mhz.

The PE control resides in the processor's I/O space at addresses 0 and 2. I/O port 0 is used to deliver data to the PE and port 2 is used to set/examine the PEBOS and examine the state of PERDY and of the control state machine. The programmer has direct control over the PEBOS signal by way of port 2: by writing a value with bit 1 set to port 2, the programmer turns on the PEBOS. Likewise, by writing a value with bit 1 clear to I/O port 2, the programmer can turn off the BOS. Using these two features allows the host to issue a BOS pulse, which is necessary to tell the PE that a new screen is coming. After setting BOS, the host can write screen data to I/O port 0 to deliver the screen data. The host writes a byte (8 pixels) at a time, and the pixels (bits) contained in that byte will be shifted serially to the PE by the PE Controller

Two mechanisms exist for writing data to this data port: direct CPU I/O writes and DMA. For direct CPU I/O writes, the CPU will read a byte from the screen image in memory and write that byte to I/O port 0. Likewise, for DMA, the DMA controller will read a byte from memory and perform an I/O write cycle to port 0. The DMA, however, has a request signal, DREQ, that must be asserted before the transfer begins. The DMA is programmed with a requester, a target, and a byte count. Here, the requester is I/O port 0 (PE data port), the target is the screen image in memory, and the byte count is $201,600 / 8 = 25,200$. Once the DMA channel is enabled, it will wait for the DREQ signal to be asserted. When DREQ is active, the DMA will read a byte from memory and write it to I/O port 0, then wait for DREQ again. When DREQ goes active again, the DMA will send the second byte, and so on. Hence, the PE controller must also handle the assertion of DREQ.

The PE controller manages this and behaves as follows. In the idle state, the controller is waiting to receive data, and DREQ is asserted in this state to tell the DMA controller that data should be delivered. When an I/O write cycle to I/O port 0 (initiated either by the CPU or by the DMA controller) occurs, the machine transitions to the RECV state. The processor asserts the PE Controller chip-select (/CS2) when there is I/O access to ports 0 or 2, so the Controller must examine bit 1 of the bus address to determine whether the access is to port 0 or port 2; the state machine only activates during writes to I/O port 0.

The controller remains in the RECV state until the I/O write is complete. The end of the I/O write cycle (denoted by the processor asserting /READY) latches the data byte on the bus into an 8-bit register in the PE Controller and sends the state machine into the SHIFT state. Also, since the internal buffer is now full, the controller turns off the DREQ signal until the buffers free, telling the DMA to stop sending bytes. At the same time the counter is cleared and starts the counting. The least significant bit (LSB) of this register is attached to the PEDATA line. When the controller enters this state, it causes the PE data shifting to activate. The controller remains in this state until the shift is done, denoted by the internal counter.

Once activated, the Shifter begins to serially deliver the data byte to the PE. Before sending the first bit, it waits for the PE to be ready for data (indicated by an active PERDY). When the PE is ready, the PECLK is asserted, which will deliver one bit to the PE (since bit 0 of the data byte is tied to PEDATA, this bit is the one that is sent to the PE). This process repeats until all 8 bits have been delivered to the PE. Once this is done, the counter generates SHIFT_DONE signal and the Shifter returns to the idle state, and awaits another byte.

Hence, the PE Controller and Data Shifter act in tandem to deliver bytes from the CPU (or DMA) to the PE display adapter. Additionally, the transfer to the PE occurs in the background; once the CPU writes the byte to the PE Controller, it can continue processing while the byte is serially sent to the PE. An AHDL description of the PE Controller design is given in Example 8.2.

Example 8.2 PE Controller.

```
INCLUDE  "8shift"; %library designs not shown here%
INCLUDE "freqdiv";
```

```
INCLUDE "4count";

SUBDESIGN PEC
(
        DATA[7..0], /CS2, A1, /READY, IOWR, PERDY : INPUT;
        clk, reset :INPUT;
        DREQ, PEDATA, PECLK :OUTPUT;
)
VARIABLE
        load_data, strobe, shift, shift_done          :NODE;
        pk                                    :NODE; %8 MHz clock%
        shifter                                       : 8shift;
        counter                      : 4count;
        fdiv                                 : freqdiv;
        ss        :MACHINE WITH STATES (idle,recv, load, shift);

BEGIN
        ss.clk=clk;
        ss.reset=reset;

        fdiv.clk=clk;
        pk=fdiv.q; %output frequency from the divider%
        PECLK=pk & shift & PERDY;
        strobe=(!/CS2)&IOWR&(!A1);
        counter.clrn=!DREQ;
        counter.clk=PECLK;
        shifter.d[7..0]=DATA[7..0];
        shifter.clk=!PECLK; %ensure bit0 is delivered before
shifting%
        shifter.ld=!(load_data&(!/READY));
        PEDATA=shifter.q;
        shift_done=counter.qd;

        TABLE
        ss, strobe, /READY, shift_done=>ss, DREQ, load_data,
shift;
        idle, 0,x,x   =>idle, 1,0,0;
        idle, 1,x,x =>recv, 1,0,0;
        recv, x,1,x =>recv, 0,0,0;
        recv, x,0,x =>load, 0,0,0;
        load, x,x,x =>shift,0,1,0;
        shift, x,x,0 =>shift, 0,0,1;
        shift, x,x,1 =>idle, 0,0,1;
```

```
        END TABLE;
END;
```

These two subsystems, the P-SRAM Controller and the PE Controller, comprise the heart of the electronics, aside from the processing core. Implementing the memory and PE controllers quickly and easily allows for reduction in the complexity of the system. A simple MAX 7000 device was chosen to implement the function of these interfaces. The simplest low-power 7032S chip accommodates both interfaces with less than 80% of its utilization. In addition, the FPLDs support 5 volt as well as 3.3 volt signals, which accommodates the 5-volt PE nicely.

The complexity of the above interfaces requires much effort to implement using standard parts or custom logic. Using FPLD, however, allows the developer to deal with a high-level description of the subsystem's behavior, rather than with cumbersome low-level details. For example, the state machine can be intuitively represented as a collection of states and transitions. Hence, mapping the memory controller and the PE interface, the two most complex blocks of logic in the VuMan 3 system, to the FPLD helped eliminate much complexity. Therefore, using the FPLD allows rapid prototyping. Without the reduction in complexity and implementation detail, these subsystems would require months to implement. With FPLD in the developer's arsenal, such logic blocks can be designed and implemented in a week instead.

8.4 Secondary Logic

An FPLD also provides additional logic signals aside from the state machines. The PCIC requires ISA-bus signals to operate properly and the FPLD is used to perform the conversion from i386EX bus to ISA bus. Namely, the FPLD provided the ISA signals of IORD (I/O read cycle), IOWR (I/O write cycle), MRD (memory read cycle), and MWR (memory write cycle). Also, the FPLD generates the system clocks, EPROM chip-select signals, and buffer control signals used in interfacing the 5 Volt PCMCIA slots to the 3.3 Volt i386EX system bus. These designs are not presented in this Chapter.

The FPLDs, coupled with the 386EX processor core, comprise the essential logic blocks. These allow the system to interface to the memory and the display adapter. The serial controller establishes communications between the CPU and the input, power, real-time clock, and serial-number subsystems. With these components interconnected, the system is ready to function.

GLOSSARY

Active-high node A node that is activated when it is assigned a value one or Vcc. In AHDL design files, an active-high node should be assigned a default value of Vcc with the Defaults statement.

Active-low node A node that is activated when it is assigned a value zero or GND. In AHDL design files, an active-low node should be assigned a default value of GND with the Defaults statement.

AHDL Altera Hardware Description Language. A design entry language which supports Boolean equation, state machine, conditional, and decode logic. It also provides access to all Altera and user-defined macrofunctions.

Antifuse Any of the programmable interconnect technologies forming electrical connection between two circuit points rather than making open connections.

Architecture Describes the behavior, dataflow, and/or structure of a VHDL entity. An architecture is created with an architecture body. A single entity can have more than one architecture. Configuration declarations are used to specify which architectures to use for each entity.

Array or group In AHDL, a group is a collection of up to 256 symbolic names that are treated as a unit. In VHDL, a group is called an array, and is not limited to 256 symbolic names.

Assignment In AHDL and VHDL, assignment refers to the transfer of a value to a symbolic name or group, usually through a Boolean equation. The value on the right side of assignment statement is assigned to the symbolic name or group on the left.

Asynchronous input An input signal that is not synchronized to the device Clock.

Back annotation The process of incorporating time delay values into a design netlist reflecting the interconnect capacitance obtained from a completed design. Also, in Altera's case, the process of copying device and resource assignments made by the compiler into an Assignment and Configuration File for a project. This process preserves the current fit in future compilations.

Buried node A combinational or registered signal that does not drive an output pin.

Buried register A register that does not drive its output to a pin. A buried register can be located on an I/O cell or on a logic cell that has no output to a pin. A buried register can be used to implement internal logic.

Cell A logic function. It may be a gate, a flip-flop, or some other structure. Usually, a cell is small compared to other circuit building blocks.

CLB Configurable Logic Block. This element is the basic building block of the Xilinx LCA product family.

Clique A group of logic functions defined as a single, named unit. The Compiler attempts to keep clique members together when it fits the project.

Clock A signal that triggers registers. In a flip-flop or state machine, the clock is an edge-sensitive signal. The output of the clock can change only on the clock edge.

Clock enable The level-sensitive signal on a flip-flop with E suffix, such as DFFE. When the Clock enable is low, Clock transitions on the Clock input of the flip-flop are ignored.

Component Specifies the ports of a primitive or macrofunction in VHDL. A component consists of the name of the primitive or macrofunction, and a list of its inputs and outputs. Components are specified in the Component declaration

Configuration It maps instances of VHDL components to design entities and describes how design entities are combined to form a complete design.

Configuration declarations are used to specify which architectures to use for each entity.

Configuration scheme The method used to load configuration data into an FPGA.

CPLD Complex Programmable Logic Device. CPLDs include an array of functionally complete or universal logic cells in an interconnection framework that has foldback connection to central programming regions.

Design file A file that contains descriptions of the logic for a project and is compiled by the compiler.

Design library Stores VHDL units that have already been compiled. These units can be referenced in VHDL designs. Design libraries can contain one or more of the following units: Entity declarations, Architecture declarations, Configuration declarations, Package declarations, and Package body declarations.

Device assignment Assigns a user-specified block or logic functions, called a chip, to a specific Altera device.

Dual-purpose pins Pins used to configure an FPGA device that can be used as I/O pins after initialization.

EDIF Electronic Design Interchange Format. An industry standard format for the transmission of design files.

EPLD EPROM Programmable Logic Devices. A PLD that uses EPROM cells to internally configure the logic function. Also called Erasable Programmable Logic Device.

Excitation function A Boolean function that specifies logic that directs state transitions in a state machine.

Expander A section in the MAX LAB containing an array of foldback NAND functions. The expander is used to increase the logical inputs to the LAB macrocell section or to make other logic and storage functions in the LAB.

Fan-in The number of input signals that feed all the input equations of a logic cell.

FastTrack interconnect Dedicated connection paths that span the entire width and height of a FLEX 8000 device. These connection paths allow the signals to travel between all LABs in a device.

Fitting The process of making a design fit into a specific architecture. Fitting involves technology mapping, placement, optimization, and partitioning among other operations.

Floorplan The physical arrangement of functions within a design relative to the other.

FPGA Field Programmable Gate Array. An array of cells that is either functionally complete or universal within a connection framework of signal routing channels.

FPLD An integrated circuit used for implementing digital hardware that allows the end user to configure the chip to realize different designs. Configuring such a device is done using either a special programming unit or by programming "in system".

Function prototype Specifies the ports of a primitive or macrofunction in AHDL. It consists of the name of the primitive or macrofunction, and a list of its inputs and outputs in exact order in which they are used. An instance of the primitive or macrofunction can be inserted with an Instance declaration or an in-line reference.

Functional simulation A simulation mode that allows the simulation of the logical performance of a project without timing information.

Functionally complete A property of some Boolean logic functions permitting them to make any logic function by using only that function. The properties include making the AND function with an invert or the OR function with an invert.

Fuse A metallic interconnect point that can be electrically changed from a short circuit to an open circuit by applying an electrical current.

Gate An electronic structure, built from transistors, that performs a function.

Gate array Array of transistors interconnected to form gates. The gates in turn are configured to form larger functions.

Gated clock A clock configuration in which the output of an AND or OR gate drives a clock.

Glitch A signal value pulse that occurs when a logic level changes two or more times over a short period.

Global signal A signal from a dedicated input pin that does not pass through the logic array before performing its specified function. Clock, Preset, Clear, and Output Enable signals can be global signals.

GND A low-level input voltage. It is the default inactive node value.

Group or array In AHDL, a group is a collection of up to 256 symbolic names that are treated as a unit. In VHDL, a group is called an array, and is not limited to 256 symbolic names.

Hard macro A function larger than a single gate but made from gates. The macro performance is invariant from placement point of view.

Include file An ASCII file that can be imported into a text design file by an AHDL Include statement. The file replaces the Include statement that calls it. Include files usually contain Function prototype or Constant statements.

Input vectors Time-ordered binary numbers representing input values sequences to a simulation program.

Instance The use of a primitive or macrofunction in a design file.

I/O cell register A register on the periphery of a FLEX 8000 device or a fast input-type logic cell that is associated with an I/O pin in some FLEX 7000 devices.

I/O feedback Feedback from the output pin on an Altera device that allows an output pin to be also used as an input pin.

LAB Logic Array Block. A LAB is the basic building block of the Altera MAX family. Each LAB contains at least one macrocell, an I/O block, and an expander product term array.

Latch A level-sensitive clocked storage unit that stores a single bit of data. A high-to-low transition on the Latch Enable signal fixes the contents of the latch at the value of the data input until the next low-to-high transition on Latch Enable.

Latch Enable A level-sensitive signal that controls a latch. When it is high, the input flows through the output; when it is low, the output holds its last value.

Logic element A basic building block of an Altera FLEX 8000 device. It consists of a look-up table (that is a function generator that quickly computes any function of four variables) and a programmable flip-flop to support sequential functions.

Long line The mechanism inside an LCA where a signal is passed through repeating amplifier to drive a larger interconnect line. Long lines are less sensitive to metal delays.

Macro A cell configuration that can be repeated as needed. It can be Hard and Soft macro.

Macrocell The portion of the FPGA that is smallest indivisible building block. In MAX devices it consists of two parts: combinatorial logic and a configurable register.

MAX Multiple Array MatriX, which is an Altera product family. It is usually considered to be a CPLD.

MAX+PLUS II Multiple Array MatriX Programmable Logic User System II. A set of tools that allow design and implementation of custom logic circuits withe Altera's MAX and FLEX devices.

Model A representation that behaves similarly to the operation of some digital circuit.

MPGA Mask-Programmable Gate Array.

MPLD Mask- Programmable Logic Device.

Netlist An ASCII file that describes a design. Minimal requirements are identification of function elements, inputs and outputs and connections.

Netlist synthesis Process of deriving a netlist from an abstract representation, usually from a hardware description language.

NRE Non-Recurring Engineering expense. It reefers to one-time charge covering the use of design facilities, masks, and overhead for test development.

One Hot Encoding A design technique used more with FPGAs than CPLDs. It assigns a single flip-flop to hold a logical one representing a state, with the rest of flip-flops being held at zeros.

Package A collection of commonly used VHDL constructs that can be shared by more than one design unit.

PAL Programmable Array Logic. A relatively small FPLD containing a programmable AND plane followed by a fixed-OR plane.

Partitioning Setting boundaries within functions of a system.

PLA Programmable Logic Array. A relatively small FPLD that contains two levels of programmable logic - an AND plane and an OR plane.

Placement The physical assignment of a logical function to a specific location within an FPGA. Once the logic function is placed, its interconnection is made by routing.

PLD Programmable Logic Device. This class of devices comprise PALs, PLAs, FPGAs and CPLDs.

Port A symbolic name that represents an input or output of a primitive or of a macrofunction design file.

Primitive One of the basic functional blocks used to design circuits with Max+PLUS II software. Primitives include buffers, flip-flops, latch, logical operators, ports, etc.

Programmable switch a user programmable switch that can connect a logic element or input/output element to an interconnect wire or one interconnect wire to another.

Project A project consists of all files that are associated with a particular design, including all subdesign files and ancillary files created by the user or by Max+PLUS II software. The project name is the same as the name of the top-level design file without the extension.

Propagation delay The time required for any signal transition to travel between pins and/or nodes in a device.

Resource A resource is a portion of a device that performs a specific, user-defined task (such as pins and logic cells).

Retargeting A process of translating a design from one FPGA or other technology to another. Retargeting involves technology mapping and optimization.

Routing Process of interconnecting previously placed logic functions.

Semicustom A general category of integrated circuits that can be configured directly by the user of the IC. It includes gate array, PLD, FPGA, PROM and EPROM devices.

Simulation The process of modeling a logical design and its stimuli in which the simulator calculates output signal models.

Slew rate The maximum time rate of voltage change without. Some FPGAs permit a fast or slow slew rate to be programmed for an output pin.

Soft micro A logic function made in a number of ways. The exact arrangement of cells may be different from one version to another. This can result in performance differences.

Speed performance The maximum speed of a circuit implemented in an FPLD. It is set by the longest delay through any path for combinational circuits and by the maximum clock frequency at which the circuit operates properly for sequential circuits.

Spike A signal value pulse that occurs when a logic level changes two or more times over a short period.

Technology mapping The process of translating the function of a design from one technology to another. All versions of the design would have the same function, but the cell used would be very different.

Universal logic cell A logic cell capable of forming any combinational logic function of the number of inputs to the cell. RAM, ROM and multiplexers have been used to form universal logic cells. Sometimes they are also called look-up tables or function generators.

Usable gates A term used to denote that not all gates on an FPGA may be accessible and used for application purposes.

Vcc A high-level input voltage represented as a high (1) logic level in binary group values. It is the default active node value in AHDL.

VHDL The VHSIC (Very High Speed Integrated Circuits) Hardware Description Language. VHDL is used to describe function, interconnect and modeling.

SELECTED READING

Due to the large amount of literature in the area of Field-Programmable Logic and digital systems design, we only suggest some of the readings.

Bolton M. *Digital Systems Design with Programmable Logic*, Addison-Wesley Publishing Co., 1990.

Brown S. et al., *Field-Programmable Gate Arrays*, Kluwer Academic Publishers, 1992.

Brown S., and Rose J.. "FPGA and CPLD Architectures: A Tutorial", IEEE Design and Test of Computers, Summer 1996.

Perry D. *VHDL*, Second Edition, McGraw-Hill, 1994.

Rose J., El Gamal A., and Sangiovanni-Vincentelli A. "Architecture of Field-Programmable Gate Arrays", Proc. IEEE, Vol. 81, No.7, July 1993.

Salcic Z. "SimP- A Simple Custom-Configurable Processor Implemented in FPGA", Tech. Report no.567/96, Auckland University, Department of Electrical and Electronic Engineering, July 1996.

Salcic Z. Maunder B., "SimP - a Core for FPLD-based Custom-Configurable Processors", Proceedings of International Conference on ASICS - ASICON '96, Shanghai, 1996.

Salcic Z., Maunder B. "CCSimP - An Instruction-Level Custom-Configurable Processor for FPLDs", Field-Programmable Logic 96, R.Hartenstein, M.Glesner (Eds), Lecture Notes in Computer Science 1142, Springer, 1996.

Smailagic, A., et. al. "Benchmarking an Interdisciplinary Concurrent Design Methodology for Electronic/Mechanical Systems" Proc. ACM/IEEE Design Automation Conference, June 1995 San Francisco, CA. 514-519.

Smailagic, A., Siewiorek, D.P. "A Case Study in Embedded System Design: The VuMan2 Wearable Computer", IEEE Design and Test of Computer, Vol. 10, No. 3, 1993; 56-67.

Smailagic, A., Siewiorek, D.P. "The CMU Mobile Computers and Their Application For Maintenance", Mobile Computing, Eds. T. Imielinski and H. Korth, Kluwer Academic Publishers, January 1996.

Smailagic,A., Siewiorek, D.P. "Interacting with CMU Wearable Computers", IEEE Personal Communications, Vol.3, No.1, Feb. 1996; 14-25.

Trimberger S., ed. *Field-Programmable Gate Array Technology*, Kluwer Academic Publishers, 1994.

Proc. IEEE Symposium FPGAs for Custom-Computing Machines, IEEE Computer Society Press, Los Alamitos. 1993-1996.

Proc. Field-Programmable Logic, FPL '94 Prague. Springer Verlag, 1994.

Field-Programmable Logic, FPL '96 Darmstadt. Springer Verlag, 1996.

Rauer, A., Smailagic, A., "VuMan 3 Hardware/ Software Design" to be submitted for publication, EDRC, CMU, 1996.

Max+PLUS II Programmable Logic Development System: AHDL, Altera Corporation, 1995.

VHDL - Language Reference, IEEE Press, 1994.

Various data sheets, application notes and application briefs by Altera Co. and Xilinx Co.

INDEX

About the Accompanying CD-ROM

Digital Systems Design and Prototyping Using Field Programmable Logic, First Edition includes a CD-ROM that contains Altera's MAX+PLUS II 7.21 Student Edition programmable logic development software. MAX+PLUS II is a fully integrated design environment that offers unmatched flexibility and performance. The intuitive graphical interface is complemented by complete and instantly accessible on-line documentation, which makes learning and using MAX+PLUS II quick and easy. MAX+PLUS II version 7.21 Student Edition offers the following features:

✓ Operates on PCs running Windows 3.1, Windows 95, and Windows NT 3.51 and 4.0.
✓ Graphical and text-based design entry, including Altera Hardware Description Language (AHDL) and VHDL.
✓ Design compilation for product-term (MAX 7000S) and look-up table (FLEX 10K) device architectures.
✓ Design verification with full timing simulation.

Installing with Windows 3.1 and Windows NT 3.51

Insert the MAX+PLUS II CD-ROM in your CD-ROM drive. In the Windows Program Manager, choose **Run** and type: *<CD-ROM drive>*: \pc\maxplus2\install in the *Command Line* box. You are guided through the installation procedure.

Installing with Windows 95 and Windows NT 4.0

Insert the MAX+PLUS II CD-ROM in your CD-ROM drive. In the Start menu, choose **Run** and type: *<CD-ROM drive>*: \pc\maxplus2\install in the *Open* box. You are guided through the installation procedure.

Registration & Additional Information

To register and obtain an authorization code to use the MAX+PLUS II software, to **http://www.altera.com/maxplus2-student**. For complete installation instructions, refer to the **read.me** file on the CD-ROM or to the *MAX+PLUS II Getting Started Manual*, available on the Altera world-wide web site (**http://www.altera.com**).